T0181836

Probability Theory and Stochastic Modelling

Volume 80

The **Probability Theory and Stochastic Modelling** series is a merger and continuation of Springer's two well established series Stochastic Modelling and Applied Probability and Probability and Its Applications series. It publishes research monographs that make a significant contribution to probability theory or an applications domain in which advanced probability methods are fundamental. Books in this series are expected to follow rigorous mathematical standards, while also displaying the expository quality necessary to make them useful and accessible to advanced students as well as researchers. The series covers all aspects of modern probability theory including

- Gaussian processes
- Markov processes
- Random fields, point processes and random sets
- Random matrices
- Statistical mechanics and random media
- Stochastic analysis

as well as applications that include (but are not restricted to):

- Branching processes and other models of population growth
- Communications and processing networks
- Computational methods in probability and stochastic processes, including simulation
- Genetics and other stochastic models in biology and the life sciences
- Information theory, signal processing, and image synthesis
- Mathematical economics and finance
- Statistical methods (e.g. empirical processes, MCMC)
- Statistics for stochastic processes
- Stochastic control
- Stochastic models in operations research and stochastic optimization
- Stochastic models in the physical sciences

More information about this series at http://www.springer.com/series/13205

Emmanuel Rio

Asymptotic Theory of Weakly Dependent Random Processes

 Springer

Emmanuel Rio
Laboratoire de Mathématiques de Versailles
Université de Versailles
Saint-Quentin-en-Yvelines
Versailles
France

ISSN 2199-3130   ISSN 2199-3149   (electronic)
Probability Theory and Stochastic Modelling
ISBN 978-3-662-57191-0   ISBN 978-3-662-54323-8   (eBook)
DOI 10.1007/978-3-662-54323-8

Mathematics Subject Classification (2010): 60-01, 60F05, 60F15, 60F17, 60E15, 60G10, 60J10, 62G07

Translation from the French language edition: *Théorie asymptotique des processus al
dépendants* by Emmanuel Rio, © Springer-Verlag Berlin Heidelberg 2000. All Rights Reserved *éatoirement faiblement*
© Springer-Verlag GmbH Germany 2017
Softcover reprint of the hardcover 1st edition 2017

Printed on acid-free paper

This Springer imprint is published by Springer Nature
The registered company is Springer-Verlag GmbH Germany
The registered company address is: Heidelberger Platz 3, 14197 Berlin, Germany

Preface

These lecture notes are the second version of the book "*Théorie asymptotique des processus aléatoires faiblement dépendants*", written in French. In the process of translation, some misprints and inaccuracies have been removed, some proofs rewritten in more detail, some recent references have been added and three new sections incorporated. However, the numeration of the initial sections remains unchanged. Below I give comments on the three newsections.

Section 1.7 gives another covariance inequality for strongly mixing sequences. This covariance inequality is an improvement of Inequality (7) of Exercise 8, obtained by the author at the end of 1991. Section 4.4 deals with the central limit theorem for triangular arrays. This section is based on a paper of the author on the Lindeberg method. In Sect. 5.6 a meaningful coupling lemma of Peligrad for unbounded random variables is stated and proved.

The mathematics of the initial version was completed in January 1999. Since that time, there has been a huge amount of new results. In particular, it is now clear that the notions of weak dependence used in this book are too restrictive for some applications, for instance in the case of Markov chains associated to some dynamical systems. Consequently some new notions of weak dependence have been introduced. We refer to Dedecker et al. (2007) for an introduction to these new notions of dependence and their associated coefficients of dependence, as well as some applications of these new techniques. We will not treat this much broader spectrum of dependence in this second edition. Finally, it is a pleasure to thank Marina Reizakis and all the staff at Springer who contributed towards the production of this book.

Versailles, France   Emmanuel Rio
September 2016

Preface to the French edition

Ce cours est une extension d'un cours commun avec Paul Doukhan effectué de 1994 à 1996 dans le cadre du DEA de modélisation stochastique et de statistiques de la faculté d'Orsay et du minicours que j'ai donné dans le cadre du séminaire Paris-Berlin 1997 de statistique mathématique. Il s'adresse avant tout aux étudiants de troisième cycle ainsi qu'aux chercheurs désireux d'appronfondir leurs connaissances sur la théorie de la addition des variables aléatoires faiblement dépendantes.

Alors que la théorie de l'addition des variables aléatoires indépendantes est désormais largement avancée, peu d'ouvrages traitent des suites de variables faiblement dépendantes. Pour ne pas nous restreindre aux chaînes de Markov, nous étudions ici les suites de variables faiblement dépendantes dites fortement mélangeantes au sens de Rosenblatt (1956) ou absolument régulières au sens de Volkonskii et Rozanov (1959). Ces suites englobent de nombreux modèles utilisés en statistique mathématique ou en économétrie, comme le montre Doukhan (1994) dans son ouvrage sur le mélange. Nous avons choisi de nous concentrer sur les inégalités de moments ou de moyennes déviations pour les sommes de variables faiblement dépendantes et sur leurs applications aux théorèmes limites.

Je voudrais remercier ici tous ceux qui m'ont aidé lors de la rédaction de cet ouvrage, en particulier Paul Doukhan et Abdelkader Mokkadem, pour toutes les connaissances qu'ils m'ont transmises ainsi que pour leurs conseils avisés. Une partie de leurs résultats est développée dans ces notes. Je voudrais également remercier Sana Louhichi et Jérôme Dedecker pour les modifications et les nombreuses améliorations qu'il mont suggérées.

Orsay, France
Emmanuel Rio
Janvier 2000

Contents

1 The Variance of Partial Sums 1
 1.1   Introduction 1
 1.2   Stationary Processes 1
 1.3   A Covariance Inequality Under Strong Mixing. 3
 1.4   Variance of Partial Sums of a Strongly Mixing Sequence 8
 1.5   Applications to Density Estimation. 15
 1.6   A Covariance Inequality Under Absolute Regularity 22
 1.7   Other Covariance Inequalities Under Strong Mixing * 26
 Exercises. 30

2 Algebraic Moments, Elementary Exponential Inequalities 33
 2.1   Introduction 33
 2.2   An Upper Bound for the Fourth Moment of Sums. 33
 2.3   Even Algebraic Moments 37
 2.4   Exponential Inequalities 40
 2.5   New Moment Inequalities. 45
 Exercises. 48

3 Maximal Inequalities and Strong Laws 51
 3.1   Introduction 51
 3.2   An Extension of the Maximal Inequality of Kolmogorov. 51
 3.3   Rates of Convergence in the Strong Law of Large Numbers 54
 Exercises. 61

4 Central Limit Theorems 65
 4.1   Introduction 65
 4.2   A Central Limit Theorem for Strongly Mixing and Stationary
     Sequences. 65
 4.3   A Functional Central Limit Theorem for the Partial
     Sum Process. 68

4.4   A Central Limit Theorem for Strongly Mixing Triangular
       Arrays*   70
Exercises   84

5   **Mixing and Coupling**   89
    5.1   Introduction   89
    5.2   A Coupling Lemma for Real-Valued Random Variables   89
    5.3   A Coupling Lemma for β-Mixing Random Variables   91
    5.4   Comparison of α-Mixing and β-Mixing Coefficients
           for Finite σ-Fields.   95
    5.5   Maximal Coupling and Absolutely Regular Sequences.   97
    5.6   An Extension of Lemma 5.2 to Unbounded Random
           Variables *   97
    Exercises   99

6   **Fuk–Nagaev Inequalities, Applications**   101
    6.1   Introduction   101
    6.2   Exponential Inequalities for Partial Sums   101
    6.3   Fuk–Nagaev Inequalities for Partial Sums   103
    6.4   Application to Moment Inequalities   106
    6.5   Application to the Bounded Law of the Iterated Logarithm   109
    Exercises   110

7   **Empirical Distribution Functions**   113
    7.1   Introduction   113
    7.2   An Elementary Estimate   114
    7.3   Functional Central Limit Theorems   116
    7.4   A Functional Central Limit Theorem for the Empirical
           Distribution Function   118
    7.5   Multivariate Distribution Functions.   121

8   **Empirical Processes Indexed by Classes of Functions**   125
    8.1   Introduction   125
    8.2   Classes of Regular Functions   126
    8.3   Maximal Coupling and Entropy with Bracketing   130
    Exercises   145

9   **Irreducible Markov Chains**   149
    9.1   Introduction   149
    9.2   Irreducible Markov Chains   150
    9.3   The Renewal Process of an Irreducible Chain   152
    9.4   Mixing Properties of Positively Recurrent Markov Chains:
           An Example   155
    9.5   Small Sets, Absolute Regularity and Strong Mixing.   159

9.6   Rates of Strong Mixing and Rates of Ergodicity of Irreducible
      Chains 164
9.7   On the Optimality of Theorem 4.2 167
Exercises... 169

Annex A: Young Duality and Orlicz Spaces 171

**Annex B: Exponential Inequalities for Sums of Independent Random
            Variables**.. 175

Annex C: Upper Bounds for the Weighted Moments............... 183

Annex D: Two Versions of a Lemma of Pisier 187

Annex E: Classical Results on Measurability 191

Annex F: The Conditional Quantile Transformation 193

Annex G: Technical Tools 195

References 197

Index 203

Notation

$a \wedge b$	$\min(a,b)$
$a \vee b$	$\max(a,b)$
$\xi.x$	Euclidean scalar product of ξ and x
x^+, x_+	For a real x, the number $\max(0,x)$
x^-, x_-	For a real x, the number $\max(0,-x)$
$[x]$	For a real x, the integer part of x
$f(x-0)$	For a function f, the left limit of f at point x
$f(x+0)$	For a function f, the right limit of f at point x
f^{-1}	The generalized inverse function of f
α^{-1}	The function defined by $\alpha^{-1}(u) = \sum_{i \in \mathbb{N}} \mathrm{II}_{u < \alpha_i}$
Q_X	For a random variable X, the inverse of $t \to \mathbb{P}(|X| > t)$
$\mathbb{E}(X)$	For an integrable random variable X, the expectation of X
B^c	The complement of B
II_B	The indicator function of B
$\mathbb{E}(X|\mathcal{A})$	Conditional expectation of X conditionally to the σ-field $\mathcal{A}$
$\mathbb{P}(B|\mathcal{A})$	Conditional expectation of II_B conditionally to $\mathcal{A}$
$\mathrm{Var}\,X$	The variance of X
$\mathrm{Cov}\,(X,Y)$	Covariance between X and Y
$|\mu|$	Measure of total variation associated to the measure μ
$\|\mu\|$	Total variation of the measure μ
$\mu \otimes \nu$	Tensor product of the measures μ and ν
$\mathcal{A} \otimes \mathcal{B}$	Tensor product of the σ-fields $\mathcal{A}$ and $\mathcal{B}$
$\mathcal{A} \vee \mathcal{B}$	The σ-field generated by $\mathcal{A} \cup \mathcal{B}$
$\sigma(X)$	σ-field generated by X
L^r	For $r \geq 1$, the space of random variables X such that $\mathbb{E}(|X|^r) < \infty$
$\|X\|_r$	For $r \geq 1$, the usual norm on L^r
L^∞	The space of almost surely bounded random variables
$\|X\|_\infty$	The usual norm on L^∞

$L^r(P)$ For a P law on $\mathcal{X}$, the space of functions f such that $\int_{\mathcal{X}} |f|^r dP < \infty$

L^ϕ The Orlicz space associated to a convex function ϕ

$\|\cdot\|_\phi$ The usual norm on L^ϕ

ϕ^* For a function ϕ, the Young dual of ϕ

Introduction

This book is a translation and a second version of "*Théorie asymptotique des processus aléatoires faiblements dépendants*", published in 2000 by Springer. It is devoted to inequalities and limit theorems for weakly dependent sequences. Our aim is to give efficient technical tools to Mathematicians or Statisticians who are interested in weak dependence. We will essentially consider classical notions of weak dependence, called mixing conditions. Sometimes we will give more general results. Nevertheless, the strong mixing coefficients of Rosenblatt (1956) will be used in most of the results.

Here the strong mixing coefficient between two σ-fields $\mathcal{A}$ and $\mathcal{B}$ is defined by

$$\alpha(\mathcal{A}, \mathcal{B}) = 2 \sup\{\mathbb{P}(A \cap B) - \mathbb{P}(A)\mathbb{P}(B) : (A, B) \in \mathcal{A} \times \mathcal{B}\}.$$

This coefficient is equal to the strong mixing coefficient of Rosenblatt (1956), up to the multiplicative factor 2. This coefficient is a measure of the dependence between $\mathcal{A}$ and $\mathcal{B}$. For example, $\alpha(\mathcal{A}, \mathcal{B}) = 0$ if and only if $\mathcal{A}$ and $\mathcal{B}$ are independent. For a sequence $(X_i)_{i \in \mathbb{Z}}$ of random variables in some Polish space $\mathcal{X}$, let $\mathcal{F}_k = \sigma(X_i : i \leq k)$ and $\mathcal{G}_l = \sigma(X_i : i \geq l)$. The strong mixing coefficients $(\alpha_n)_{n \geq 0}$ of the sequence $(X_i)_{i \in \mathbb{Z}}$ are defined by

$$\alpha_0 = 1/2 \text{ and } \alpha_n = \sup_{k \in \mathbb{Z}} \alpha(\mathcal{F}_k, \mathcal{G}_{k+n}) \quad \text{for any } n > 0. \tag{1}$$

The sequence $(X_i)_{i \in \mathbb{Z}}$ is said to be strongly mixing in the sense of Rosenblatt (1956) if $\lim_{n \uparrow \infty} \alpha_n = 0$. In the stationary case, this means that the σ-field $\mathcal{G}_n$ of the future after time n is asymptotically independent of $\mathcal{F}_0$, which is the σ-field of the past before time 0. We refer to Bradley (1986, 2005) for other coefficients of weak dependence and the relations between them.

In these notes, we will mainly establish results for strongly mixing sequences or for absolutely regular sequences in the sense of Volkonskii and Rozanov (1959). Indeed, these notions of weak dependence are less restrictive than the notions of

ρ-mixing and uniform mixing in the sense of Ibragimov (1962). For example, in the case of autoregressive models with values in $\mathbb{R}^d$ defined by the recursive equation

$$X_{n+1} = f(X_n) + \varepsilon_{n+1}, \tag{2}$$

for some sequence of independent and identically distributed integrable innovations $(\varepsilon_n)_n$ with a positive continuous bounded density, the stationary sequence $(X_i)_{i \in \mathbb{Z}}$ solution of (2) is uniformly mixing in the sense of Ibragimov only if the function f is uniformly bounded over $\mathbb{R}^d$. This condition is too restrictive for the applications. By contrast, the stationary solution of (2) is strongly mixing with a geometric rate of strong mixing as soon as there exists $M > 0, s > 0$ and $\rho < 1$ such that

$$\mathbb{E}(|f(x) + \varepsilon_0|^s) \leq \rho |x|^s \quad \text{for } x > M \text{ and } \sup_{|x| \leq M} \mathbb{E}(|f(x) + \varepsilon_0|^s) < \infty. \tag{3}$$

We refer to Doukhan and Ghindès (1983) and to Mokkadem (1985) for more about the model (2), and to Doukhan (1994) for other examples of Markov models satisfying mixing conditions. Although the notions of strong mixing or absolute regularity are less restrictive than the notions of ρ-mixing and uniform mixing, they are adequate for the applications. For example, Viennet (1997) obtains optimal results for linear estimators of the density in the case of absolutely regular sequences.

We now summarize the contents of these lecture notes. Our main tools are covariance inequalities for random variables satisfying mixing conditions and coupling results which are similar to the coupling theorems of Berbee (1979) and Goldstein (1979). Chapters 1–4 are devoted to covariance inequalities, moment inequalities and classical limit theorems. Chapters 5–8 mainly use coupling techniques. The coupling techniques are applied to the law of the iterated logarithm for partial sums in Chap. 6 and then to empirical processes in Chaps. 7 and 8.

In Chap. 1, we give covariance inequalities for random variables satisfying a strong mixing condition or an absolute regularity condition. Let us recall Ibragimov's (1962) covariance inequality for bounded random variables: if X and Y are uniformly bounded real-valued random variables, then

$$|\text{Cov}(X, Y)| \leq 2\alpha(\sigma(X), \sigma(Y)) \|X\|_\infty \|Y\|_\infty, \tag{4}$$

where $\sigma(X)$ and $\sigma(Y)$ denote the σ-fields generated by X and Y, respectively. We give extensions of (4) to unbounded random variables. We then apply these covariance inequalities to get estimates of the variance of partial sums. In the dependent case, the variance of the sum may be much larger than the sum of variances. We refer to Bradley (1997) for lower bounds for the variance of partial sums in the strong mixing case. Nevertheless, adequate applications of the variance estimates still provide efficient results. For example, we give in Sects. 1.5 and 1.6 some impressive applications of mixing conditions to density estimation. In Sect. 1.7* we give other covariance inequalities (the * indicates that the section is new to this edition).

Chapter 2 is devoted to the applications of covariance inequalities to moment inequalities for partial sums. In Sects. 2.2 and 2.3, we apply the covariance inequalities of Chap. 1 to algebraic moments of sums. Our methods are similar to the methods proposed by Doukhan and Portal (1983, 1987). They lead to Rosenthal type inequalities. In Sects. 2.4 and 2.5, we prove Marcinkiewicz type moment inequalities for the absolute moments of order $p > 2$, and we give a way to derive exponential inequalities from these results. In Chap. 3 we give extensions of the maximal inequalities of Doob and Kolmogorov to dependent sequences. These maximal inequalities are then used to obtain Baum–Katz type laws of large numbers, and consequently rates of convergence in the strong law of large numbers. We also derive moment inequalities of order p for p in $]1, 2[$ from these inequalities.

Chapter 4 is devoted to the classical central limit theorem for partial sums of random variables. We start by considering strictly stationary sequences. In this context, we apply projective criteria derived from Gordin's martingale approximation theorem (1969) to get the central limit theorem for partial sums of a strongly mixing sequence. We then give a uniform functional central limit theorem in the sense of Donsker for the normalized partial sum process associated to a stationary and strongly mixing sequence. The proof of tightness is based on the maximal inequalities of Chap. 3. Section 4.4*, at the end of the chapter, is devoted to the central limit theorem for triangular arrays.

In Chap. 5, we give coupling results for weakly dependent sequences, under assumptions of strong mixing or β-mixing. In particular, we recall and we prove Berbee's coupling Lemma (1979), which characterizes the β-mixing coefficient between a σ-field $\mathcal{A}$ and the σ-field $\sigma(X)$ generated by some random variable X with values in some Polish space. If $(\Omega, \mathcal{T}, \mathbb{P})$ contains an auxiliary atomless random variable independent of $\mathcal{A} \vee \sigma(X)$, then one can construct a random variable X^* with the same law as X, independent of $\mathcal{A}$ and such that

$$\mathbb{P}(X = X^*) = 1 - \beta(\mathcal{A}, \sigma(X)). \qquad (5)$$

We give a constructive proof of (5) for random variables with values in $[0, 1]$. This proof is more technical than the usual proof. Nevertheless, the constructive proof is more informative than the usual proof. In particular, using a comparison theorem between α-mixing coefficients and β-mixing coefficients for purely atomic σ-fields due to Bradley (1983), one can obtain (see Exercise 1) the following upper bound for the so-constructed random variables:

$$\mathbb{E}(|X - X^*|) \leq 4\alpha(\mathcal{A}, \sigma(X)). \qquad (6)$$

In Sect. 5.2, we give a direct proof of (6) with an improved constant. Our method of proof is based on the conditional quantile transformation. Section 5.6* contains an extension of (6) to unbounded random variables.

Chapters 6–8 are devoted to the applications of these coupling results. In Chap. 6, we prove that Inequality (6) yields efficient deviation inequalities for partial sums of real-valued random variables. In particular, we generalize the Fuk–Nagaev deviation inequalities (1971) to partial sums of strongly mixing sequences of real-valued random variables. For example, for sums $S_k = X_1 + \cdots + X_k$ of real-valued and centered random variables X_i satisfying $\|X_i\|_\infty \leq 1$, we prove that, for any $\lambda > 0$ and any $r \geq 1$,

$$\mathbb{P}\left(\sup_{k \in [1,n]} |S_k| \geq 4\lambda \right) \leq 4\left(\left(1 + \frac{\lambda^2}{rs_n^2}\right)^{-r/2} + \frac{n\alpha_{[\lambda/r]}}{\lambda} \right), \tag{7}$$

where

$$s_n^2 = \sum_{i=1}^{n} \sum_{j=1}^{n} |\operatorname{Cov}(X_i, X_j)|.$$

This inequality is an extension of the Fuk–Nagaev inequality to weakly dependent sequences. Theorem 6.2 provides an extension in the general case of unbounded random variables. Choosing $r = 2 \log \log n$, we then apply (7) to the bounded law of the iterated logarithm. In Chaps. 7 and 8, we apply (5)–(7) to empirical processes associated to dependent observations. We refer the reader to Dudley (1984) and to Pollard (1990) for more about the theory of functional limit theorems for empirical processes. In Chap. 7, we give uniform functional central limit theorems for the normalized and centered empirical distribution function associated to real-valued random variables or to random variables with values in $\mathbb{R}^d$. We prove that the uniform functional central limit theorem for the normalized and centered empirical distribution function holds true under the strong mixing condition $\alpha_n = O(n^{-1-\varepsilon})$ for any $d > 1$. The strong mixing condition does not depend on the dimension, in contrast to the previous results. The proof is based on Inequality (7). This inequality does not provide uniform functional central limit theorems for empirical processes indexed by large classes of sets. For this reason, we give a more general result in Chap. 8, which extends Dudley's theorem (1978) for empirical processes indexed by classes of sets to β-mixing sequences. The proof of this result is based on the maximal coupling theorem of Goldstein (1979).

Chapter 9, which concludes these lecture notes, is devoted to the mixing properties of irreducible Markov chains and the links between ergodicity, return times, absolute regularity and strong mixing. We also prove the optimality of some of the results of the previous chapters on various examples of Markov chains. The Annexes are devoted to convex analysis, exponential inequalities for sums of independent random variables, tools for empirical processes, upper bounds for the weighted moments introduced in Chaps. 1 and 2, measurability questions and quantile transformations.

Chapter 1
The Variance of Partial Sums

1.1 Introduction

In order to study the deviation and the limiting distribution of a partial sum of real-valued random variables, one of the main steps is to study the variance of this sum. For independent random variables, the variance of the sum is the sum of individual variances. This assertion is generally false for dependent random variables, with the notable exception of martingale difference sequences. However, for stationary sequences, the so-called series of covariances provides asymptotic estimates of the variance of partial sums. Consequently, for dependent sequences, one needs to give conditions on the sequence implying the convergence of this series. Such conditions are given, for example, by the so-called mixing assumptions. In this chapter, we start by giving classical results on the variance of partial sums in the stationary case. Next we give bounds on the covariance between two random variables under a strong mixing condition on these random variables. These results are then applied to the variance of partial sums of strongly mixing sequences. Then we give applications to integrated risks of kernel density estimators or linear estimators of the density, under mixing assumptions. The end of this section is devoted to the so-called β-mixing sequences, applications of this notion to density estimation and to another covariance inequality in the strong mixing case.

1.2 Stationary Processes

In this section we recall some basic results on partial sums of random variables in the stationary case. We start by recalling the definitions of strict stationarity, and stationarity at second order.

Définition 1.1 Let $T = \mathbb{Z}$ or $T = \mathbb{N}$. The process $(X_t)_{t \in T}$ is said to be strictly stationary if, for any positive integer t and any finite subset S of T,

© Springer-Verlag GmbH Germany 2017
E. Rio, *Asymptotic Theory of Weakly Dependent Random Processes*,
Probability Theory and Stochastic Modelling 80,
DOI 10.1007/978-3-662-54323-8_1

$$\{X_{s+t} : s \in S\} \text{ has the same distribution as } \{X_s : s \in S\}. \tag{1.1}$$

For sequences $(X_t)_{t \in T}$ of real-valued and square-integrable random variables, $(X_t)_{t \in T}$ is said to be stationary at second order if, for any positive integer t and any (u, v) in $T \times T$,

$$\mathbb{E}(X_u) = \mathbb{E}(X_v) \text{ and } \mathbb{E}(X_{t+u} X_{t+v}) = \mathbb{E}(X_u X_v). \tag{1.2}$$

We now define the covariance between two real-valued integrable random variables X and Y such that XY is still integrable by

$$\text{Cov}(X, Y) = \mathbb{E}(XY) - \mathbb{E}(X)\mathbb{E}(Y). \tag{1.3}$$

Throughout, we assume that the random variables X_t take their values in $\mathbb{R}$. Assume now that $(X_t)_{t \in T}$ is stationary at second order. Let

$$S_n = X_1 + \cdots + X_n, \ V_n = \text{Var } S_n \text{ and } v_n = V_n - V_{n-1}, \tag{1.4}$$

with the conventions that $S_0 = 0$ and $V_0 = 0$. Clearly $V_n = v_1 + \cdots + v_n$. We now estimate v_k. From the bilinearity and the symmetry of the covariance

$$v_k = \text{Cov}(S_k, S_k) - \text{Cov}(S_{k-1}, S_{k-1}) = \text{Var } X_k + 2 \sum_{i=1}^{k-1} \text{Cov}(X_i, X_k).$$

Hence, for second order stationary sequences,

$$v_k = \text{Var } X_0 + 2 \sum_{i=1}^{k-1} \text{Cov}(X_0, X_i) \tag{1.5}$$

and

$$V_n = n \text{Var } X_0 + 2 \sum_{i=1}^{n} (n - i) \text{Cov}(X_0, X_i). \tag{1.6}$$

From (1.5) and (1.6) we get the elementary lemma below.

Lemma 1.1 *Let $(X_i)_{i \in \mathbb{N}}$ be a sequence of real-valued random variables, stationary at second order. Assume that the so-called series of covariances*

$$\text{Var } X_0 + 2 \sum_{i=1}^{\infty} \text{Cov}(X_0, X_i)$$

converges. Then the sum v of this series is nonnegative, and $n^{-1} \text{Var } S_n$ converges to v.

Remark 1.1 Sometimes $v = 0$. For example, if there exists some stationary sequence $(Y_i)_{i \in \mathbb{N}}$ such that $X_i = Y_i - Y_{i-1}$, satisfying the condition $\lim_{n \to +\infty} \mathrm{Cov}(Y_0, Y_n) = 0$, then $v = 0$. In this specific case, the sequence $(S_n)_{n>0}$ is bounded in L^2, and consequently in probability.

Proof of Lemma 1.1 Since $V_n/n = (v_1 + \cdots + v_n)/n$, the convergence of v_k to v implies that of (V_n/n) to v via the Césaro mean theorem. Furthermore, $v \geq 0$ since $(V_n/n) \geq 0$ for any positive integer n. ■

We now give a sufficient condition to ensure the convergence of the above series, which will be called throughout the series of covariances of $(X_i)_{i \in \mathbb{N}}$.

Lemma 1.2 *Let $(X_i)_{i \in \mathbb{N}}$ be a sequence of real-valued random variables, stationary at second order. Assume that there exists a sequence of nonnegative reals $(\delta_i)_{i \geq 0}$ such that*

$$\mathrm{Cov}(X_0, X_i) \leq \delta_i \ \text{for any } i \geq 0 \ \text{and} \ \Delta = \delta_0 + 2 \sum_{i>0} \delta_i < \infty. \qquad (i)$$

Then $\mathrm{Var} \, X_0 + 2 \sum_{i=1}^{\infty} \mathrm{Cov}(X_0, X_i)$ converges to an element v of $[0, \Delta]$. Furthermore,

$$\mathrm{Var} \, S_n \leq n\delta_0 + 2 \sum_{i=1}^{n} (n - i)\delta_i \leq n\Delta \ \text{and} \ v_k \leq \delta_0 + 2 \sum_{i=1}^{k-1} \delta_i. \qquad (1.7)$$

Proof of Lemma 1.2 Write

$$\mathrm{Cov}(X_0, X_i) = \delta_i - (\delta_i - \mathrm{Cov}(X_0, X_i)).$$

The series $\sum_i (\delta_i - \mathrm{Cov}(X_0, X_i))$ is a series of nonnegative reals and, consequently, converges in $\bar{\mathbb{R}}^+$. Hence the series of covariances converges to v in $[-\infty, \Delta]$. By the Césaro mean theorem, $n^{-1}\mathrm{Var} \, S_n$ converges to v. It follows that v belongs to $[0, \Delta]$. Now (1.7) holds due to the nonnegativity of the numbers δ_i. ■

1.3   A Covariance Inequality Under Strong Mixing

In this section, we give a covariance inequality for real-valued random variables satisfying a strong mixing condition. This inequality will be applied in Sect. 1.4 to obtain conditions on the strong mixing coefficients of stationary sequences ensuring the convergence of series of covariances.

We start by defining the strong mixing coefficient of Rosenblatt (1956) between two σ-fields $\mathcal{A}$ and $\mathcal{B}$ of $(\Omega, \mathcal{T}, \mathbb{P})$. We refer to Bradley (1986, 2005) for more about strong mixing conditions and weak dependence coefficients and to Bradley (2007) for a much more extensive treatment. In order to get more general results, we will also give less restrictive coefficients associated to real-valued random variables.

For real-valued random variables X and Y, we set

$$\alpha(X, Y) = 2 \sup_{(x,y) \in \mathbb{R}^2} |\mathbb{P}(X > x, Y > y) - \mathbb{P}(X > x)\mathbb{P}(Y > y)|, \qquad (1.8a)$$

and then

$$\alpha(\mathcal{A}, Y) = \sup_{A \in \mathcal{A}} \alpha(\mathbb{1}_A, Y). \qquad (1.8b)$$

Note that $\alpha(X, Y) = 0$ means that X and Y are independent. Also $\alpha(\mathcal{A}, Y) = 0$ if and only if Y is independent of $\mathcal{A}$. The strong mixing coefficient between two σ-fields $\mathcal{A}$ and $\mathcal{B}$ is defined by

$$\alpha(\mathcal{A}, \mathcal{B}) = \sup_{B \in \mathcal{B}} \alpha(\mathcal{A}, \mathbb{1}_B) = 2 \sup\{|\text{Cov}(\mathbb{1}_A, \mathbb{1}_B)| : (A, B) \in \mathcal{A} \times \mathcal{B}\}. \qquad (1.8c)$$

This coefficient is the Rosenblatt strong mixing coefficient, up to the multiplicative factor 2. This coefficient vanishes if and only if the σ-fields are independent. Now, by the Cauchy–Schwarz inequality,

$$|\text{Cov}(\mathbb{1}_A, \mathbb{1}_B)| \leq \sqrt{\text{Var } \mathbb{1}_A \text{Var } \mathbb{1}_B} \leq 1/4.$$

It follows that

$$0 \leq \alpha(\mathcal{A}, \mathcal{B}) \leq 1/2. \qquad (1.9)$$

Furthermore, $\alpha(\mathcal{A}, \mathcal{B}) = 1/2$ if and only if there exists some event A in $\mathcal{A} \cap \mathcal{B}$ with $\mathbb{P}(A) = 1/2$. In a similar way the coefficients defined in (1.8a) and (1.8b) are each bounded above by $1/2$. Let us now give a slightly different formulation of these coefficients. Clearly

$$\alpha(\mathcal{A}, \mathcal{B}) = \sup\{|\text{Cov}(\mathbb{1}_A - \mathbb{1}_{A^c}, \mathbb{1}_B)| : (A, B) \in \mathcal{A} \times \mathcal{B}\}. \qquad (1.10a)$$

Next

$$\text{Cov}(\mathbb{1}_A - \mathbb{1}_{A^c}, \mathbb{1}_B) = \mathbb{E}((\mathbb{P}(B \mid \mathcal{A}) - \mathbb{P}(B))(\mathbb{1}_A - \mathbb{1}_{A^c}))$$

and consequently, for a fixed B, the maximum over $\mathcal{A}$ is attained by the measurable set $A = (\mathbb{P}(B \mid \mathcal{A}) > \mathbb{P}(B))$. Consequently,

$$\alpha(\mathcal{A}, \mathcal{B}) = \sup\{\mathbb{E}(|\mathbb{P}(B \mid \mathcal{A}) - \mathbb{P}(B)|) : B \in \mathcal{B}\}. \qquad (1.10b)$$

In the same way one can prove that

$$\alpha(\mathcal{A}, X) = \sup_{x \in \mathbb{R}} \mathbb{E}(|\mathbb{P}(X \leq x \mid \mathcal{A}) - \mathbb{P}(X \leq x)|). \qquad (1.10c)$$

In order to state the covariance inequality, which takes into account the marginal distribution of the random variables, we now introduce more notation.

Notation 1.1 For any nonincreasing and càdlàg function f with domain the interval I, let f^{-1} denote the càdlàg inverse function of f, which is defined by

$$f^{-1}(u) = \inf\{x \in I : f(x) \le u\}.$$

The basic property of f^{-1} is that:

$$x < f^{-1}(u) \text{ if and only if } f(x) > u.$$

If f is a nondecreasing and càdlàg function, $f^{-1}(u)$ will be the infimum of the set of reals x in I such that $f(x) \ge u$. In that case, the inverse is left-continuous and

$$x \ge f^{-1}(u) \text{ if and only if } f(x) \ge u.$$

The distribution function F of a real-valued random variable X is defined by $F(x) = \mathbb{P}(X \le x)$. This function is nondecreasing and left-continuous. The quantile function of $|X|$, which is the inverse of the nonincreasing and left-continuous tail function of $|X|$, $H_X(t) = \mathbb{P}(|X| > t)$, is denoted by Q_X. For any monotonous function f, we set

$$f(x - 0) = \lim_{y \nearrow x} f(y) \text{ and } f(x + 0) = \lim_{y \searrow x} f(y).$$

Theorem 1.1 *Let X and Y be integrable real-valued random variables. Assume that XY is integrable and let $\alpha = \alpha(X, Y)$ be defined by (1.8a). Then*

$$|\mathrm{Cov}(X, Y)| \le 2 \int_0^\alpha Q_X(u) Q_Y(u) du \le 4 \int_0^{\alpha/2} Q_X(u) Q_Y(u) du. \qquad (a)$$

Conversely, for any symmetric distribution functions F and G and any α in $[0, 1/2]$, one can construct random variables X and Y with respective distribution functions F and G such that $\alpha(\sigma(X), \sigma(Y)) \le \alpha$ and

$$\mathrm{Cov}(X, Y) \ge \int_0^{\alpha/2} Q_X(u) Q_Y(u) du, \qquad (b)$$

provided that $Q_X Q_Y$ is integrable on $[0, 1]$.

Remark 1.2 Theorem 1.1 is due to Rio (1993). We refer to Dedecker and Doukhan (2003) for extensions of (a) and to Dedecker, Gouëzel and Merlevède (2010) for applications of (a) to Markov chains associated to intermittent maps (these chains fail to be strongly mixing in the sense of Rosenblatt). If $\alpha = 1/2$ (no mixing constraint), Theorem 1.1(a) ensures that

$$|\text{Cov}(X, Y)| \leq 2 \int_0^{1/2} Q_X(u)Q_Y(u)du. \tag{1.11a}$$

Now, if (Z, T) is a pair of random variables with the same marginal distributions as (X, Y), then

$$\mathbb{E}(|ZT|) \leq \int_0^1 Q_X(u)Q_Y(u)du. \tag{1.11b}$$

If, furthermore, X and Y have symmetric distributions, then the upper bound in (1.11b) is attained for $Z = \varepsilon Q_X(U)$ and $T = \varepsilon Q_Y(U)$, where U is uniformly distributed in $[0, 1]$ and ε is a symmetric sign, independent of U; see Fréchet (1951, 1957), Bass (1955) or Bártfai (1970). Consequently, up to a constant factor, (a) cannot be improved.

Let us give another byproduct of the covariance inequality. Let $\mathcal{A}$ be a σ-field of $(\Omega, \mathcal{T}, \mathbb{P})$ and X be a real-valued random variable. Let Y be a real-valued random variable with mean 0, and $\alpha = \alpha(\mathcal{A}, Y)$. Let $\varepsilon_{\mathcal{A}}$ be the random variable defined by $\varepsilon_{\mathcal{A}} = 1$ if $\mathbb{E}(Y \mid \mathcal{A}) > 0$ and $\varepsilon_{\mathcal{A}} = -1$ otherwise. Then, from Theorem 1.1(a),

$$\mathbb{E}(|X\mathbb{E}(Y \mid \mathcal{A})|) = \text{Cov}(\varepsilon_{\mathcal{A}}X, Y) \leq 2 \int_0^{\alpha} Q_X(u)Q_Y(u)du. \tag{1.11c}$$

Note that, if U has the uniform law over $[0, 1]$, $Q_X(U)$ has the same law as $|X|$. Hence, if $|X|$ and $|Y|$ are almost surely bounded, then (a) implies that

$$|\text{Cov}(X, Y)| \leq 2\alpha \|X\|_{\infty} \|Y\|_{\infty}, \tag{1.12a}$$

which gives again the covariance inequality of Ibragimov (1962).

For unbounded random variables, Hölder's inequality applied to the upper bound in (a) proves that if p, q and r are strictly positive reals such that $p^{-1} + q^{-1} + r^{-1} = 1$, then

$$|\text{Cov}(X, Y)| \leq 2\alpha^{1/p} \|X\|_q \|Y\|_r, \tag{1.12b}$$

which provides a new constant in Inequality (2.2) of Davydov (1968).

Under the weaker tail conditions

$$\mathbb{P}(|X| > x) \leq (\Lambda_q(X)/x)^q \text{ and } \mathbb{P}(|Y| > y) \leq (\Lambda_r(Y)/y)^r,$$

Theorem 1.1(a) gives:

$$|\text{Cov}(X, Y)| \leq 2p\alpha^{1/p}\Lambda_q(X)\Lambda_r(Y). \tag{1.12c}$$

Consequently, one can obtain the same dependence in α as in (1.12b) under weaker conditions on the tails of the random variables. In the next section, we will prove that Theorem 1.1(a) provides more efficient upper bounds on the variance of partial sums than (1.12b).

Proof of Theorem 1.1 We first prove (a). Let

$$X^+ = \sup(0, X) \text{ and } X^- = \sup(0, -X).$$

Clearly

$$X = X^+ - X^- = \int_0^{+\infty} (\mathrm{1\!I}_{X>x} - \mathrm{1\!I}_{X<-x})dx. \tag{1.13}$$

Writing Y in the same manner and applying Fubini's Theorem, we get that

$$\mathrm{Cov}(X, Y) = \int_0^\infty \int_0^\infty \mathrm{Cov}(\mathrm{1\!I}_{X>x} - \mathrm{1\!I}_{X<-x}, \mathrm{1\!I}_{Y>y} - \mathrm{1\!I}_{Y<-y})dxdy. \tag{1.14}$$

In order to bound $|\mathrm{Cov}(X, Y)|$, we now prove that:

$$|\mathrm{Cov}(\mathrm{1\!I}_{X>x} - \mathrm{1\!I}_{X<-x}, \mathrm{1\!I}_{Y>y} - \mathrm{1\!I}_{Y<-y})| \leq 2\inf(\alpha, \mathbb{P}(|X| > x), \mathbb{P}(|Y| > y)). \tag{1.15}$$

Obviously the term on the left-hand side is bounded above by 2α. Since X and Y play a symmetric role, it only remains to prove that the term on the left-hand side is bounded above by $2\mathbb{P}(|X| > x)$. From the elementary inequality $|\mathrm{Cov}(S, T)| \leq 2\|S\|_1\|T\|_\infty$ applied to $S = \mathrm{1\!I}_{X>x} - \mathrm{1\!I}_{X<-x}$ and $T = \mathrm{1\!I}_{Y>y} - \mathrm{1\!I}_{Y<-y}$, we infer that

$$|\mathrm{Cov}(\mathrm{1\!I}_{X>x} - \mathrm{1\!I}_{X<-x}, \mathrm{1\!I}_{Y>y} - \mathrm{1\!I}_{Y<-y})| \leq 2\mathbb{P}(|X| > x),$$

which completes the proof of (1.15). From (1.15) and (1.14), we have

$$|\mathrm{Cov}(X, Y)| \leq 2\int_0^\infty \int_0^\infty \inf(\alpha, \mathbb{P}(|X| > x), \mathbb{P}(|Y| > y))dxdy. \tag{1.16}$$

Now

$$\inf(\alpha, \mathbb{P}(|X| > x), \mathbb{P}(|Y| > y)) = \int_0^\alpha \mathrm{1\!I}_{u<\mathbb{P}(|X|>x)}\,\mathrm{1\!I}_{u<\mathbb{P}(|Y|>y)}du.$$

Since $(u < \mathbb{P}(|X| > x))$ if and only if $(x < Q_X(u))$, one can write (1.16) as follows

$$|\mathrm{Cov}(X, Y)| \leq 2\int_0^\infty \int_0^\infty \left(\int_0^\alpha \mathrm{1\!I}_{x<Q_X(u)}\,\mathrm{1\!I}_{y<Q_Y(u)}du\right)dxdy. \tag{1.17}$$

To complete the proof of (a), it is then enough to apply Fubini's Theorem.

To prove (b), we construct a pair (U, V) of random variables with marginal distributions the uniform law over $[0, 1]$, satisfying $\alpha(\sigma(U), \sigma(V)) \leq \alpha$ and such that (b) holds true for $(X, Y) = (F^{-1}(U), G^{-1}(V))$.

Let a be any real in $[0, 1]$, and (Z, T) be a random variable with the uniform distribution over $[0, 1] \times [a/2, 1 - a/2]$. Set

$$(U, V) = 1\!\!1_{Z \in [a/2, 1-a/2]}(Z, T) + 1\!\!1_{(Z \notin [a/2, 1-a/2[}(Z, Z). \tag{1.18}$$

Then the random variables U and V are uniformly distributed over $[0, 1]$. We now prove that

$$\alpha(\sigma(U), \sigma(V)) \le 2a. \tag{1.19}$$

Let $P_{U,V}$ denote the law of (U, V) and P_U, P_V denote the laws of U and V. Clearly

$$\|P_{U,V} - P_U \otimes P_V\| = 4a - 2a^2$$

(here $\| \cdot \|$ denotes the total variation of the signed measure). Now, by (1.10b) and Remark 1.4 in Sect. 1.6, the total variation of $P_{U,V} - P_U \otimes P_V$ is greater than 2α. Hence (1.19) holds true.

Next, let $(X, Y) = (F^{-1}(U), G^{-1}(V))$. Since X is a measurable function of U and Y is a measurable function of V, $\alpha(\sigma(X), \sigma(Y)) \le \alpha$. Now

$$XY = F^{-1}(Z)G^{-1}(Z)1\!\!1_{Z \notin [a/2, 1-a/2]} + F^{-1}(Z)G^{-1}(T)1\!\!1_{Z \in [a/2, 1-a/2]}.$$

Taking the expectation of this formula (recall that Z and T are independent and that $\mathbb{E}(G^{-1}(T)) = 0$), we get that

$$\mathbb{E}(XY) = \int_0^{a/2} F^{-1}(u)G^{-1}(u)du + \int_{1-a/2}^1 F^{-1}(u)G^{-1}(u)du.$$

Next, from the symmetry of F,

$$F^{-1}(1 - u) = -F^{-1}(u) = Q_X(2u) \text{ almost everywhere on } [0, 1/2]$$

(with the same equality for G). Hence

$$\text{Cov}(X, Y) \ge 2 \int_0^{a/2} Q_X(2u)Q_Y(2u)du = \int_0^a Q_X(u)Q_Y(u)du,$$

which completes the proof of (b). ∎

1.4   Variance of Partial Sums of a Strongly Mixing Sequence

In this section, we apply Theorem 1.1 to get upper bounds on the variance of $S_n = X_1 + \cdots + X_n$, in the strong mixing case. In this section, the sequence of strong mixing coefficients $(\alpha_n)_{n \ge 0}$ of $(X_i)_{i \in \mathbb{N}}$ is defined by

$$\alpha_0 = 1/2 \text{ and } \alpha_n = \sup_{\substack{(i,j) \in \mathbb{N}^2 \\ |i-j| \ge n}} \alpha(\sigma(X_i), \sigma(X_j)) \text{ for } n > 0. \tag{1.20}$$

By definition the so-defined sequence is nonincreasing.

For x in $\mathbb{R}$, let $\alpha(x) = \alpha_{[x]}$, the square brackets designating the integer part. Let

$$\alpha^{-1}(u) = \inf\{k \in \mathbb{N} : \alpha_k \leq u\} = \sum_{i \geq 0} \mathbb{1}_{u < \alpha_i} \qquad (1.21)$$

(the second equality is due to the monotonicity properties of $(\alpha_i)_{i \geq 0}$). Starting from Theorem 1.1(a), we now get an upper bound on the variance of partial sums.

Corollary 1.1 *Let $(X_i)_{i \in \mathbb{N}}$ be a sequence of real-valued random variables. Set $Q_k = Q_{X_k}$. Then*

$$\text{Var } S_n \leq \sum_{i=1}^{n} \sum_{j=1}^{n} |\text{Cov}(X_i, X_j)| \leq 4 \sum_{k=1}^{n} \int_0^1 [\alpha^{-1}(u) \wedge n] Q_k^2(u) du. \qquad (a)$$

In particular, setting

$$M_{2,\alpha}(Q) = \int_0^1 \alpha^{-1}(u) Q^2(u) du$$

for any nonnegative and nonincreasing function Q from $[0, 1]$ into $\mathbb{R}$, we have:

$$\text{Var } S_n \leq 4 \sum_{k=1}^{n} M_{2,\alpha}(Q_k). \qquad (b)$$

Proof (b) is an immediate consequence of (a). To prove (a), notice that

$$\alpha^{-1}(u) \wedge n = \sum_{i=0}^{n-1} \mathbb{1}_{u < \alpha_i}.$$

Clearly

$$\text{Var } S_n \leq \sum_{(i,j) \in [1,n]^2} |\text{Cov}(X_i, X_j)|. \qquad (1.22)$$

Now, by Theorem 1.1(a),

$$|\text{Cov}(X_i, X_j)| \leq 2 \int_0^{\alpha_{|i-j|}} Q_i(u) Q_j(u) du \leq \int_0^{\alpha_{|i-j|}} (Q_i^2(u) + Q_j^2(u)) du.$$

Hence

$$\sum_{(i,j)\in[1,n]^2} |\text{Cov}(X_i, X_j)| \le 2 \sum_{i=1}^{n} \int_0^1 \sum_{j=1}^{n} \text{1\!I}_{u<\alpha_{|i-j|}} Q_i^2(u)du$$

$$\le 4 \sum_{i=1}^{n} \int_0^1 [\alpha^{-1}(u) \wedge n] Q_i^2(u)du. \qquad (1.23)$$

Both (1.22) and (1.23) imply Corollary 1.1(a).   ∎

We now apply Theorem 1.1 and Corollary 1.1 to stationary sequences.

Corollary 1.2 *Let $(X_i)_{i\in\mathbb{N}}$ be a strictly stationary sequence of real-valued random variables. Then*

$$|\text{Cov}(X_0, X_i)| \le 2 \int_0^{\alpha_i} Q_0^2(u)du. \qquad (a)$$

Consequently the series of covariances $\text{Var } X_0 + 2\sum_{i>0} \text{Cov}(X_0, X_i)$ *converges to a finite nonnegative real σ^2 as soon as*

$$M_{2,\alpha}(Q_0) = \int_0^1 \alpha^{-1}(u) Q_0^2(u)du < +\infty. \qquad (DMR)$$

In that case

$$\text{Var } S_n \le 4n M_{2,\alpha}(Q_0), \ \lim_{n\uparrow\infty} n^{-1} \text{Var } S_n = \sigma^2 \ \text{and } \sigma^2 \le 4M_{2,\alpha}(Q_0). \qquad (b)$$

Proof Inequality (a) is an immediate consequence of Theorem 1.1. Now, starting from (a), we prove Corollary 1.2. Let $\delta_i = 2\int_0^{\alpha_i} Q_0^2(u)du$. Clearly the sequence $(\delta_i)_i$ satisfies condition (i) of Lemma 1.2, provided that (DMR) holds true. Corollary 1.2 then follows from Lemmas 1.1 and 1.2.   ∎

   In some sense, in the strong mixing case the weighted moments $M_{2,\alpha}(Q_k)$ play the same role as the usual second moments in the independent case. In the section below, we give upper bounds for these weighted moments under various conditions on the tails of the random variables and on the strong mixing coefficients.

Upper Bounds for the Weighted Moments $M_{2,\alpha}$

First, note that, if U is a random variable with uniform law over $[0, 1]$, then $Q_k^2(U)$ has the same law as X_k^2.
   If the sequence $(X_i)_{i\in\mathbb{N}}$ is m-dependent,

$$\alpha^{-1}(u) = \sum_{i=0}^{m} \text{1\!I}_{u<\alpha_i} \le m+1,$$

which entails that $M_{2,\alpha}(Q_k) \leq (m+1)\mathbb{E}(X_k^2)$. In that case condition (DMR) holds as soon as X_0^2 is integrable.

If the random variables $|X_k|$ are uniformly bounded by some positive constant M, then $Q_k^2(u) \leq M^2$ and

$$M_{2,\alpha}(Q_k) \leq M^2 \int_0^1 \alpha^{-1}(u)du \leq M^2 \int_0^\infty \alpha(x)dx.$$

Then condition (DMR) holds if and only if

$$\sum_{i \geq 0} \alpha_i < \infty, \tag{1.24}$$

which is the classical condition of Ibragimov.

We now give a condition on the tail distribution of the random variables X_k. Assume that, for some $r > 2$, $\mathbb{P}(|X_k| > x) \leq (c/x)^r$ for any positive x and any integer k. Then the quantile functions Q_k are bounded above by $cu^{-1/r}$, whence

$$M_{2,\alpha}(Q_k) \leq c^2 \sum_{i=0}^\infty \int_0^{\alpha_i} u^{-2/r}du \leq \frac{c^2 r}{r-2} \sum_{i \geq 0} \alpha_i^{1-2/r}.$$

Consequently condition (DMR) holds as soon as

$$\sum_{i \geq 0} \alpha_i^{1-2/r} < \infty. \tag{IBR}$$

In the stationary case, Ibragimov (1962) obtains the convergence of the series of covariances under (IBR) together with the more restrictive assumption of existence of the moment of order r for the random variables X_i.

Assume now that, for some $r > 2$, the random variables X_k belong to L^r. Then, by Hölder's inequality,

$$M_{2,\alpha}(Q_k) \leq \left(\int_0^1 [\alpha^{-1}(u)]^{r/(r-2)} du \right)^{1-2/r} \left(\int_0^1 Q_k^r(u) du \right)^{2/r}.$$

The second integral on the right-hand side is equal to $\|X_k\|_r^2$, since $Q_k(U)$ has the same distribution as $|X_k|$. Now let $[y]$ denote the integer part of y and set $\alpha(y) = \alpha_{[y]}$. Since the inverse function of $u \to [\alpha^{-1}(u)]^{r/(r-2)}$ is $x \to \alpha(x^{1-2/r})$,

$$\int_0^1 [\alpha^{-1}(u/2)]^{r/(r-2)} du = \int_0^\infty \alpha(x^{1-2/r}) dx$$

$$= \sum_{i \geq 0} ((i+1)^{r/(r-2)} - i^{r/(r-2)})\alpha_i.$$

Now

$$(i + 1)^{r/(r-2)} - i^{r/(r-2)} \leq r(r - 2)^{-1}(i + 1)^{2/(r-2)},$$

which entails that

$$\int_0^1 [\alpha^{-1}(u)]^{r/(r-2)} du \leq \frac{r}{r - 2} \sum_{i \geq 0}(i + 1)^{2/(r-2)} \alpha_i.$$

Hence

$$M_{2,\alpha}(Q_k) \leq \exp(2/r)\left(\sum_{i \geq 0}(i + 1)^{2/(r-2)} \alpha_i\right)^{1-2/r} \|X_k\|_r^2. \tag{1.25a}$$

In particular, in the stationary case, condition (DMR) holds if

$$\sum_{i \geq 0}(i + 1)^{2/(r-2)} \alpha_i < \infty. \tag{1.25b}$$

Under the same moment condition, Davydov's (1968) covariance inequality ensures the convergence of the series of covariances under the more restrictive condition (IBR). For example, if

$$\alpha_k = O(k^{-r/(r-2)}(\log k)^{-\theta})$$

(note that $r/(r - 2)$ is the critical exponent), (1.25b) holds for $\theta > 1$ and (IBR) needs the stronger condition $\theta > r/(r - 2)$.

In order to give conditions ensuring (DMR) under more general moment conditions on the random variables X_k and on $\alpha^{-1}(U)$, we now introduce the class of convex functions

$$\Phi = \{\phi : \mathbb{R}^+ \to \mathbb{R}^+ : \phi \text{ convex, nondecreasing, } \phi(0) = 0, \lim_{+\infty} \frac{\phi(x)}{x} = \infty\}. \tag{1.26}$$

For any ϕ in Φ, the Young dual function ϕ^* is defined by

$$\phi^*(y) = \sup_{x>0}(xy - \phi(x)).$$

We refer to Annex A for some properties of this involutive transformation and to (A.5), annex A, for a definition of the Orlicz norms below. Inequality (A.8), Annex A, ensures that

$$M_{2,\alpha}(Q_k) = \mathbb{E}(\alpha^{-1}(U)Q_k^2(U)) \leq 2\|\alpha^{-1}(U)\|_{\phi^*}\|X_k^2\|_\phi.$$

Suppose there exists a $c' > 0$ such that $\phi(X_k^2/c')$ is integrable. Then the above inequality shows that condition (DMR) is satisfied if

$$\mathbb{E}(\phi^*(\alpha^{-1}(U)/c)) < +\infty \tag{1.27}$$

for some positive constant c. Since U has the uniform law over $[0, 1]$,

$$\mathbb{P}(\alpha^{-1}(U) > x) = \mathbb{P}(U < \alpha(x)) = \alpha(x).$$

Hence, by (A.3), Annex A,

$$\begin{aligned} c\mathbb{E}(\phi^*(\alpha^{-1}(U)/c)) &= \int_0^\infty \mathbb{P}(\alpha^{-1}(U) > x)(\phi^*)'(x/c)dx \\ &= \int_0^\infty \alpha(x)\phi'^{-1}(x/c)dx, \end{aligned} \tag{1.28}$$

where ϕ'^{-1} denotes the left-continuous inverse of the derivative of ϕ. Since ϕ'^{-1} is nondecreasing, condition (DMR) is satisfied if

$$\sum_{i \geq 0} \alpha_i \phi'^{-1}((i+1)/c) < \infty \tag{1.29}$$

for some positive constant c. Bulinskii and Doukhan (1987) generalized Davydov's covariance inequality (1968) to Orlicz spaces. For sequences of random variables with a finite ϕ-moment, they obtained the convergence of the series of covariances under the summability condition

$$\sum_{i \geq 0} \phi^{-1}(1/\alpha_i)\alpha_i < \infty, \tag{HER}$$

which was introduced by Herrndorf (1985) for the central limit theorem. In Rio (1993) it is shown that this condition is more restrictive than (1.29), which we now detail for fast mixing rates.

Geometric and subgeometric rates of mixing. For $b > 0$, consider the function

$$\phi_b(x) = x(\log(1+x))^b.$$

This function belongs to Φ and has derivative

$$\phi_b'(x) = (\log(1+x))^b + bx(1+x)^{-1}(\log(1+x))^{b-1}. \tag{1.30}$$

The inverse function of ϕ_b is equivalent to $x \to \exp(x^{1/b})$ as x tends to ∞. Consequently, if

$$\mathbb{E}(X_0^2(\log(1+|X_0|))^b) < \infty, \tag{1.31}$$

then, by (1.29), condition (DMR) holds true if there exists some positive τ such that

$$\alpha_i = O(\exp(-\tau i^{1/b})) \text{ as } i \to \infty. \tag{1.32}$$

In particular, if $\alpha_i = O(a^i)$ for some a in $]0, 1[$ (geometric mixing rate) (1.32) and (1.31) hold with $b = 1$, and (DMR) holds as soon as

$$\mathbb{E}(X_0^2 \log(1 + |X_0|)) < \infty. \tag{1.33}$$

Let us compare (1.33) with condition (HER). Under (1.31), (HER) holds if and only if the series $\sum_{i \geq 0} |\log \alpha_i|^{-b}$ converges. Condition (1.32) does not ensure the convergence of this series. For example, under (1.33), (HER) does not ensure the convergence of the series of covariances for geometric rates of convergence.

Numerical comparisons. We now compare the constants arising from our covariance inequality and Davydov's inequality in the following case: the sequence $(X_i)_{i \in \mathbb{N}}$ is strictly stationary, $\mathbb{E}(X_0^4) < \infty$ and $\alpha_i \leq 2^{-1-i}$. Applying Davydov's covariance inequality with the constant in (1.12b), we get that

$$|\text{Var } S_n - n\text{Var } X_0| \leq 4n \|X_0\|_4^2 \sum_{i > 0} \sqrt{\alpha_i} \leq 2(\sqrt{2} + 2)n \|X_0\|_4^2. \tag{1.34}$$

This upper bound has to be multiplied by $2\sqrt{2}$ when using the initial constant of Davydov (1968).

Now, by Theorem 1.1(a) together with the Schwarz inequality,

$$|\text{Var } S_n - n\text{Var } X_0| \leq 4n \int_0^{\alpha_1} (\alpha^{-1}(u) - 1) Q_0^2(u) du$$
$$\leq 4n \|(\alpha^{-1}(U) - 1)_+\|_2 \|X_0\|_4^2.$$

Since the inverse function of $u \to (\alpha^{-1}(u) - 1)_+^2$ is $x \to \alpha(1 + \sqrt{x})$,

$$\|(\alpha^{-1}(U) - 1)_+\|_2^2 = \int_0^\infty \alpha(1 + \sqrt{x}) dx,$$

and our bounds lead to

$$|\text{Var } S_n - n\text{Var } X_0| \leq 2\sqrt{6}n \|X_0\|_4^2. \tag{1.35}$$

The numerical value of the constant in (1.34) is 6.83 while the numerical value of the constant in (1.35) is 4.89.

1.5 Applications to Density Estimation

In this section, $(X_i)_{i \in \mathbb{N}}$ is a strictly stationary sequence of random variables with values in $\mathbb{R}^d$. The marginal distribution P is assumed to be absolutely continuous with respect to the Lebesgue measure on $\mathbb{R}^d$. We are interested in estimating the density f of P. In this section, the strong mixing coefficients of $(X_i)_{i \in \mathbb{N}}$ are defined by (1.20).

Kernel Density Estimators

We start by defining kernel density estimators. Let $K : \mathbb{R}^d \to \mathbb{R}$ be an integrable kernel satisfying

$$\int_{\mathbb{R}^d} K(x)dx = 1 \text{ and } \int_{\mathbb{R}^d} K^2(x)dx < +\infty. \qquad (H1)$$

Let $(h_n)_{n>0}$ be a sequence of positive reals converging to 0. The kernel density estimator f_n at time n is defined by

$$f_n(x) = (nh_n^d)^{-1} \sum_{k=1}^{n} K(h_n^{-1}(x - X_k)). \qquad (1.36)$$

For stationary and strongly mixing sequences, Mokkadem (1985) proves that the L^2 norm of $f_n - \mathbb{E}(f_n)$ has the same order of magnitude as in the independent case, under condition (1.24). He also obtains some related results for L^p-norms. In this section, we recall Mokkadem's result in the case $p = 2$ and we give a proof of this result.

Theorem 1.2 *Let $(X_i)_{i \in \mathbb{N}}$ be a strictly stationary sequence of observations with values in $\mathbb{R}^d$. Let f_n be the kernel density estimator as defined in (1.36). Assume that (H1) holds. Then*

$$\int_{\mathbb{R}^d} \text{Var } f_n(x)dx \le 8(nh_n^d)^{-1} \sum_{i=0}^{n-1} \alpha_i \int_{\mathbb{R}^d} K^2(x)dx.$$

Proof Set $h_n = h$ and $K_h(x) = K(x/h)$. Let P denote the common marginal distribution of the observations X_k. Define the empirical measures P_n and the normalized and centered empirical measure Z_n by

$$P_n = n^{-1} \sum_{k=1}^{n} \delta_{X_k} \text{ and } Z_n = \sqrt{n}(P_n - P). \qquad (1.37)$$

With these notations, Theorem 1.2 is equivalent to the inequality below:

$$\int_{\mathbb{R}^d} \mathbb{E}((Z_n * K_h(x))^2) dx \leq 8 \sum_{i=0}^{n-1} \alpha_i \int_{\mathbb{R}^d} K_h^2(x) dx. \tag{1.38}$$

Now, by the Parseval–Plancherel identity,

$$\int_{\mathbb{R}^d} (Z_n * K_h(x))^2 dx = (2\pi)^{-d} \int_{\mathbb{R}^d} |\hat{Z}_n(\xi)\hat{K}_h(\xi)|^2 d\xi,$$

and consequently

$$\int_{\mathbb{R}^d} \mathbb{E}((Z_n * K_h(x))^2) dx \leq (2\pi)^{-d} \int_{\mathbb{R}^d} \mathbb{E}(|\hat{Z}_n(\xi)|^2) |\hat{K}_h(\xi)|^2 d\xi$$

$$\leq (2\pi)^{-d} \sup_{\xi \in \mathbb{R}^d} \mathbb{E}(|\hat{Z}_n(\xi)|^2) \int_{\mathbb{R}^d} |\hat{K}_h(\xi)|^2 d\xi$$

$$\leq \sup_{\xi \in \mathbb{R}^d} \mathbb{E}(|\hat{Z}_n(\xi)|^2) \int_{\mathbb{R}^d} K_h^2(x) dx$$

by the Parseval–Plancherel identity again. Next

$$n|\hat{Z}_n(\xi)|^2 = \left(\sum_{k=1}^{n} (\cos(\xi.X_k) - \mathbb{E}(\cos(\xi.X_k))) \right)^2 +$$

$$\left(\sum_{k=1}^{n} (\sin(\xi.X_k) - \mathbb{E}(\sin(\xi.X_k))) \right)^2.$$

To finish the proof of (1.38), we start from the above equality and we apply Corollary 1.1 twice. Noting that the random variables $\cos(\xi.X_k)$ and $\sin(\xi.X_k)$ take their values in $[-1, 1]$, we get that

$$\mathbb{E}(|\hat{Z}_n(\xi)|^2) \leq 8 \int_0^1 (\alpha^{-1}(u) \wedge n) du,$$

which completes the proof of (1.38).   ∎

Projection Estimators

Let $w : \mathbb{R}^d \to \mathbb{R}^+$ be a nonnegative and locally square integrable function. The space $\mathbb{R}^d$ is equipped with the measure $w(x)dx$. Let $(e_j)_{j>0}$ be a complete orthonormal system in the Hilbert space $L^2(w(x)dx)$. Suppose that the observations X_k have a common law P with density $f(x)$ with respect to the Lebesgue measure on $\mathbb{R}^d$. Assume furthermore that f belongs to the Hilbert space $L^2(w(x)dx)$. Let

$$a_j = \int_{\mathbb{R}^d} f(x)e_j(x)w(x)dx.$$

Then, by the Plancherel identity,

$$f(t) = \sum_{j>0} a_j e_j(t).$$

Let $\Pi_m f$ denote the orthogonal projection of f on the vector space generated by $e_1, \ldots, e_m$. Then

$$\Pi_m f = \sum_{j=1}^{m} a_j e_j.$$

Furthermore, $\Pi_m f$ converges to f in $L^2(w(x)dx)$ as m tends to infinity. Now define the estimators $\hat{a}_j$ of the coefficients a_j by $\hat{a}_j = P_n(we_j)$. Then $\mathbb{E}(\hat{a}_j) = a_j$ and, under suitable conditions on w and f, $\hat{a}_j$ converges to a_j as n tends to infinity. Now, we set, for some nondecreasing sequence $(m_n)_n$ of positive integers tending to infinity (to be chosen later),

$$\hat{f}_n = \sum_{j=1}^{m_n} \hat{a}_j e_j = n^{-1} \sum_{j=1}^{m_n} \sum_{k=1}^{n} w(X_k)e_j(X_k)e_j. \tag{1.39}$$

Then

$$\mathbb{E}(\hat{f}_n) = \sum_{j=1}^{m_n} a_j e_j = \Pi_{m_n} f. \tag{1.40}$$

Hence $\mathbb{E}(\hat{f}_n)$ converges to f in $L^2(w(x)dx)$ as n tends to infinity. In order to get the convergence of $\hat{f}_n$ to f in $L^2(w(x)dx)$, it is then enough to bound the variance of $\hat{f}_n$. In Theorem 1.3 below, we give an upper bound which is efficient for unconditional orthonormal bases or Riesz bases. We refer to Leblanc (1995) for wavelet estimators of the density and to Ango-Nzé (1994) for general linear estimators of the density.

Theorem 1.3 *Let* $(X_i)_{i \in \mathbb{N}}$ *be a strictly stationary sequence of observations with density* f *in the Hilbert space* $L^2(w(x)dx)$. *Then*

$$n \int_{\mathbb{R}^d} w(x) \operatorname{Var} \hat{f}_n(x) dx \le 4 \sum_{i=0}^{n-1} \alpha_i \sup_{x \in \mathbb{R}^d} \left(w(x) \sum_{j=1}^{m_n} |e_j(x)| \right)^2. \tag{a}$$

Let h *be defined by* $h(x) = (1+x)\log(1+x) - x$. *Then*

$$n \int_{\mathbb{R}^d} w(x) \operatorname{Var} \hat{f}_n(x) dx \leq 20 \|\alpha^{-1}(U) \wedge n\|_h \sup_{x \in \mathbb{R}^d} \left(w^2(x) \sum_{j=1}^{m_n} e_j^2(x) \right). \qquad (b)$$

Proof For convenience, write $m = m_n$. Since $(e_j)_{j \in [1,m]}$ is an orthonormal system, it is easy to check that

$$n \int_{\mathbb{R}^d} w(x) \operatorname{Var} \hat{f}_n(x) dx = \sum_{j=1}^{m} \operatorname{Var} Z_n(we_j). \qquad (1.41)$$

Let $\varepsilon_1, \ldots, \varepsilon_m$ be a finite Rademacher sequence, that is, a sequence of symmetric and independent signs. Suppose furthermore that this sequence is independent of $(X_i)_{i \in \mathbb{N}}$. Then

$$\sum_{j=1}^{m} \operatorname{Var} Z_n(we_j) = \mathbb{E}\left(\left(Z_n \left(\sum_{j=1}^{m} \varepsilon_j we_j \right) \right)^2 \right). \qquad (1.42)$$

We now proceed conditionally on $\varepsilon_1, \ldots, \varepsilon_m$. By Corollary 1.1,

$$\mathbb{E}\left(\left(Z_n \left(\sum_{j=1}^{m} \varepsilon_j we_j \right) \right)^2 \mid \varepsilon_1, \ldots, \varepsilon_m \right) \leq 4 \sum_{i=0}^{n-1} \alpha_i \left\| \sum_{j=1}^{m} \varepsilon_j w(X_0) e_j(X_0) \right\|_\infty^2.$$

Noting that

$$\left\| \sum_{j=1}^{m} \varepsilon_j w(X_0) e_j(X_0) \right\|_\infty \leq \left\| \sum_{j=1}^{m} w(X_0) |e_j(X_0)| \right\|_\infty,$$

we then get Theorem 1.3(a).

We now prove (b). Let

$$c = \|\alpha^{-1}(U)\|_h \quad \text{and} \quad c' = \left\| \left(\sum_{j=1}^{m} \varepsilon_j w(X_0) e_j(X_0) \right)^2 \right\|_{h^*}.$$

For $(\varepsilon_1, \ldots, \varepsilon_m)$ in $\{-1, 1\}^m$, let $Q_{\varepsilon_1, \ldots, \varepsilon_m}$ be the quantile function of the random variable $\left| \sum_{j=1}^{m} \varepsilon_j w(X_0) e_j(X_0) \right|$. By Corollary 1.1 applied conditionally on $(\varepsilon_1, \ldots, \varepsilon_m)$,

$$\mathbb{E}\left(\left(Z_n \left(\sum_{j=1}^{m} \varepsilon_j we_j \right) \right)^2 \right) \leq 2^{2-m} \sum_{(\varepsilon_1, \ldots, \varepsilon_m) \in \{-1,1\}^m} \int_0^1 [\alpha^{-1}(u) \wedge n] Q_{\varepsilon_1, \ldots, \varepsilon_m}^2(u) du.$$

$$\qquad (1.43)$$

Next, by inequality (A.7) in Annnex A, applied with $x = [\alpha^{-1}(u) \wedge n]/c$ and $y = Q_{\varepsilon_1, \ldots, \varepsilon_m}^2(u)/c'$, we have

$$\frac{1}{cc'}\mathbb{E}\left(\left(Z_n\left(\sum_{j=1}^{m}\varepsilon_j we_j\right)\right)^2\right) \le 4 + 2^{2-m}\sum_{(\varepsilon_1,\dots,\varepsilon_m)}\int_0^1 h^*(Q^2_{\varepsilon_1,\dots,\varepsilon_m}(u)/c')du. \quad (1.44)$$

Now $Q^2_Z(U)$ has the law of Z^2. Hence

$$\int_0^1 h^*(Q^2_{\varepsilon_1,\dots,\varepsilon_m}(u)/c')du = \mathbb{E}\left(h^*\left(\left|\sum_{j=1}^{m}\varepsilon_j w(X_0)e_j(X_0)\right|^2\right)\right),$$

which, together with both (1.44) and inequalities (1.41) and (1.42), ensures that

$$n\int_{\mathbb{R}^d}w(x)\mathrm{Var}\,\hat{f}_n(x)dx \le$$

$$8\|\alpha^{-1}(U)\wedge n\|_h\left\|\left(\sum_{j=1}^{m}\varepsilon_j w(X_0)e_j(X_0)\right)^2\right\|_{h^*}. \quad (1.45)$$

To complete the proof, it remains to show that

$$\left\|\left(\sum_{j=1}^{m}\varepsilon_j w(X_0)e_j(X_0)\right)^2\right\|_{h^*} \le \frac{5}{2}\sup_{x\in\mathbb{R}^d}\left(w^2(x)\sum_{j=1}^{m}e_j^2(x)\right). \quad (1.46)$$

Proof of (1.46). Let $(Y_1,\dots,Y_m)$ be a Gaussian random vector with independent and $N(0,1)$-distributed components. For any reals $p_1,\dots,p_m$ and any positive integer k,

$$\mathbb{E}((p_1\varepsilon_1+\cdots p_m\varepsilon_m)^{2k}) \le \mathbb{E}((p_1Y_1+\cdots p_mY_m)^{2k}).$$

Consequently, for any positive s such that $2s(p_1^2+\cdots p_m^2) < 1$,

$$\mathbb{E}(\exp(s(p_1\varepsilon_1+\cdots p_m\varepsilon_m)^2)) \le \mathbb{E}(\exp(s(p_1Y_1+\cdots p_mY_m)^2))$$
$$\le (1-2s(p_1^2+\cdots p_m^2))^{-1/2}. \quad (1.47)$$

Let $\psi(x) = (1-2x)^{-1/2}-1-x$. Since the Legendre transform h^* of the function h is $h^*(x) = e^x - 1 - x$ (cf. Annex A), it follows from (1.47) that

$$\mathbb{E}\left(h^*\left(s\left|\sum_{j=1}^{m}\varepsilon_j w(X_0)e_j(X_0)\right|^2\right)\mid X_0 = x\right) \le$$

$$\psi(sw^2(x)(e_1^2(x)+\cdots+e_m^2(x))), \quad (1.48)$$

provided that $sw^2(x)(e_1^2(x)+\cdots+e_m^2(x)) < 1/2$. Hence

$$\left\|\left(\sum_{j=1}^{m}\varepsilon_j w(X_0)e_j(X_0)\right)^2\right\|_{h^*} \leq \frac{1}{\psi^{-1}(1)}\sup_{x\in\mathbb{R}^d}\left(w^2(x)\sum_{j=1}^{m}e_j^2(x)\right).$$

(1.46) then follows from the fact that $\psi(2/5) \leq 1$.   ∎

Application of Theorem 1.3(a) to unconditional bases. Suppose that $(e_1, \ldots, e_m)$ is an unconditional basis, which means that there exists some positive constant K, independent of m, such that

$$\left\|\sum_{j=1}^{m}c_j we_j\right\|_{\infty} \leq K\sqrt{m}\sup_{j\in[1,m]}|c_j|.$$

(1.49)

Then Theorem 1.3 ensures that

$$\int_{\mathbb{R}^d}w(x)\text{Var }\hat{f}_n(x)dx \leq 4K^2\frac{m}{n}\sum_{i=0}^{n-1}\alpha_i.$$

(1.50)

For example, suppose that $w(x) = \text{II}_{]0,1]}$ and let us consider the histogram bases

$$e_{j,m}(x) = \sqrt{m}\,\text{II}_{](j-1)/m,j/m]} \text{ for } j \in [1,m].$$

(1.51)

Then (1.49) holds with $K = 1$.

We now apply these facts to laws with density f with support in $[0, 1]$. Let δ be some real in $]0, 1]$. We denote by $\mathcal{F}(\delta, C)$ the class of densities with support included in $[0, 1]$ such that

$$|f(x) - f(y)| \leq C|x - y|^\delta \text{ for any } (x, y) \in [0, 1]^2.$$

(1.52)

Starting from (1.52), one can easily bound the absolute value of the bias term $\mathbb{E}(f_n) - f$. Together with Theorem 1.3(a), this yields the result below on the mean integrated square error.

Corollary 1.3 *Let $(X_i)_{i\in\mathbb{N}}$ be a strictly stationary sequence of real-valued random variables. Assume that the random variable X_1 has a density f in the class $\mathcal{F}(\delta, C)$, for some $C \geq 1$. For the bases defined in (1.51), let*

$$D_2(\mathcal{F}(\delta, C)) = \inf_{m>0}\sup_{f\in\mathcal{F}(\delta,C)}\int_0^1\mathbb{E}\left(\left(f(x) - \sum_{j=1}^{m}P_n(e_{j,m})e_{j,m}(x)\right)^2\right)dx.$$

Then

$$D_2(\mathcal{F}(\delta, C)) \leq 8C^2\left(n^{-1}\sum_{k=0}^{n-1}\alpha_k\right)^{2\delta/(1+2\delta)}.$$

Consequently, if $\sum_{k\geq 0}\alpha_k < \infty$, then

$$D_2(\mathcal{F}(\delta, C)) = O(n^{-2\delta/(1+2\delta)}) \qquad (a)$$

and, if $\alpha_k = O(k^{-a})$ for some a in $]0, 1[$, then

$$D_2(\mathcal{F}(\delta, C)) = O(n^{-2a\delta/(1+2\delta)}). \qquad (b)$$

Remark 1.3 (a) gives an upper bound of the same order as in the independent case. By contrast (b) provides a slower rate. It would be interesting to study the L^p risks for $p < 2$ in that case.

Application of Theorem 1.3(b) to Riesz bases. Suppose that $(e_j)_{j>0}$ satisfies the Riesz condition:

$$\|w^2(e_1^2 + \cdots + e_m^2)\|_\infty \leq K'm \qquad (1.53)$$

for some positive constant K'. Then, by Theorem 1.3(b),

$$\int_{\mathbb{R}^d} w(x)\mathrm{Var}\ \hat{f}_n(x)dx \leq 20K'\frac{m}{n}\|\alpha^{-1}(U) \wedge n\|_h. \qquad (1.54)$$

Under the mixing condition

$$\sum_{i\geq 0} \alpha_i |\log \alpha_i| < \infty, \qquad (1.55)$$

(1.54) together with Theorem 1.3(b) yields

$$\int_{\mathbb{R}^d} w(x)\mathrm{Var}\ \hat{f}_n(x)dx = O(m/n). \qquad (1.56)$$

For example, if $m = 2m' + 1$, $w(x) = \mathbb{1}_{[0,1]}$ and $e_1, \ldots, e_m$ is the system of trigonometric polynomials of degree at most m', (1.53) holds with $K' = 1$. In that case, if $\mathcal{F}$ is the ball of radius R of the Sobolev space $H_s(T)$ on the torus $T = [0, 1]/\{0 = 1\}$, Theorem 1.3(b) yields the minimax bound

$$\inf_{m>0} \sup_{f\in\mathcal{F}} \int_0^1 \mathbb{E}\left(\left(f(x) - \sum_{j=1}^m P_n(e_j)e_j(x)\right)^2\right)dx = O(n^{-2s/(1+2s)}). \qquad (1.57)$$

Note that, for β-mixing sequences, (1.57) holds under the condition $\sum_{k\geq 0} \beta_k < \infty$, as shown by Viennet (1997).

1.6   A Covariance Inequality Under Absolute Regularity

In this section, we state and prove Delyon's covariance inequality (1990) for random variables satisfying β-mixing type conditions. We start with the definition of the β-mixing coefficient between two σ-fields. These coefficient are also called absolute regularity coefficients.

Definition 1.2 Let $\mathcal{A}$ and $\mathcal{B}$ two σ-fields of $(\Omega, \mathcal{T}, \mathbb{P})$. Let the probability measure $P_{\mathcal{A} \otimes \mathcal{B}}$ be defined on $(\Omega \times \Omega, \mathcal{A} \otimes \mathcal{B})$ as the image of $\mathbb{P}$ under the canonical injection i from $(\Omega, \mathcal{T}, \mathbb{P})$ into $(\Omega \times \Omega, \mathcal{A} \otimes \mathcal{B})$ defined by $i(\omega) = (\omega, \omega)$. Then

$$P_{\mathcal{A} \otimes \mathcal{B}}(A \times B) = \mathbb{P}(A \cap B).$$

Now, let us denote by $P_{\mathcal{A}}$ (resp. $P_{\mathcal{B}}$) the restriction of P to $\mathcal{A}$ (resp. $\mathcal{B}$). The β-mixing coefficient of Rozanov and Volkonskii (1959) is defined by

$$\beta(\mathcal{A}, \mathcal{B}) = \sup_{C \in \mathcal{A} \otimes \mathcal{B}} |P_{\mathcal{A} \otimes \mathcal{B}}(C) - P_{\mathcal{A}} \otimes P_{\mathcal{B}}(C)|.$$

This coefficient is also called the coefficient of absolute regularity.

Remark 1.4 Let $C = (A \times B) \cup (A^c \times B^c)$. Then

$$P_{\mathcal{A} \otimes \mathcal{B}}(C) - P_{\mathcal{A}} \otimes P_{\mathcal{B}}(C) = 2(\mathbb{P}(A \cap B) - \mathbb{P}(A)\mathbb{P}(B)).$$

This equality ensures that $\beta(\mathcal{A}, \mathcal{B}) \geq \alpha(\mathcal{A}, \mathcal{B})$.

We now introduce the stronger uniform mixing coefficient of Ibragimov (1962).

Definition 1.3 The uniform mixing or φ-mixing coefficient $\varphi(\mathcal{A}, \mathcal{B})$ between two σ-fields $\mathcal{A}$ and $\mathcal{B}$ is defined by

$$\varphi(\mathcal{A}, \mathcal{B}) = \sup_{\substack{(A,B) \in \mathcal{A} \times \mathcal{B} \\ \mathbb{P}(A) \neq 0}} |\mathbb{P}(B \mid A) - \mathbb{P}(B)|.$$

This coefficient belongs to $[0, 1]$. In contrast to the β-mixing coefficient, $\varphi(\mathcal{A}, \mathcal{B}) \neq \varphi(\mathcal{B}, \mathcal{A})$.

In order to compare $\beta(\mathcal{A}, \mathcal{B})$ and $\varphi(\mathcal{A}, \mathcal{B})$, we will use the following identity, whose proof is left to the reader:

$$\beta(\mathcal{A}, \mathcal{B}) = \frac{1}{2} \sup \left\{ \sum_{i \in I} \sum_{j \in J} |\mathbb{P}(A_i \cap B_j) - \mathbb{P}(A_i)\mathbb{P}(B_j)| \right\}, \qquad (1.58)$$

the maximum being taken over all finite partitions $(A_i)_{i \in I}$ and $(B_j)_{j \in J}$ of Ω with the sets A_i in $\mathcal{A}$ and the sets B_j in $\mathcal{B}$.

Now, starting from (1.58) we compare the two mixing coefficients. Fix i in I. Let J' be the set of elements j of J such that

$$\mathbb{P}(A_i \cap B_j) \geq \mathbb{P}(A_i)\mathbb{P}(B_j)$$

and let B be the union of the sets B_j for j in J'. Then

$$\frac{1}{2} \sum_{j \in J} |\mathbb{P}(A_i \cap B_j) - \mathbb{P}(A_i)\mathbb{P}(B_j)| = \mathbb{P}(A_i)(\mathbb{P}(B \mid A_i) - \mathbb{P}(B))$$

$$\leq \mathbb{P}(A_i)\varphi(\mathcal{A}, \mathcal{B}). \qquad (1.59)$$

Summing over I, we thus get that

$$\beta(\mathcal{A}, \mathcal{B}) \leq \varphi(\mathcal{A}, \mathcal{B}). \qquad (1.60)$$

Ibragimov (1962) has given a suitable covariance inequality for real-valued random variables under a uniform mixing condition. It is important to note that this inequality cannot be deduced from Theorem 1.1(a). Nevertheless, this covariance inequality is a corollary of a more powerful covariance inequality involving the β-mixing coefficient, due to Delyon (1990). We now state and prove this inequality. We refer to Dedecker (2004) for an extension to a weaker notion of dependence.

Theorem 1.4 Let $\mathcal{A}$ and $\mathcal{B}$ be two σ-fields of $(\Omega, \mathcal{T}, \mathbb{P})$. Then there exist random variables $d_{\mathcal{A}}$ and $d_{\mathcal{B}}$ with values in $[0, 1]$, respectively $\mathcal{A}$ and $\mathcal{B}$-measurable, satisfying

$$\mathbb{E}(d_{\mathcal{A}}) = \mathbb{E}(d_{\mathcal{B}}) = \beta(\mathcal{A}, \mathcal{B}),$$

and such that, for any pair (p, q) of reals in $[1, \infty]$ with $(1/p) + (1/q) = 1$ and any random vector (X, Y) in $L^p(\mathcal{A}) \times L^q(\mathcal{B})$,

$$|\mathrm{Cov}(X, Y)| \leq 2\mathbb{E}^{1/p}(d_{\mathcal{A}}|X|^p)\mathbb{E}^{1/q}(d_{\mathcal{B}}|Y|^q). \qquad (a)$$

Furthermore, $\|d_{\mathcal{A}}\|_\infty \leq \varphi(\mathcal{A}, \mathcal{B})$ and $\|d_{\mathcal{B}}\|_\infty \leq \varphi(\mathcal{B}, \mathcal{A})$. Hence

$$|\mathrm{Cov}(X, Y)| \leq 2\varphi(\mathcal{A}, \mathcal{B})^{1/p}\varphi(\mathcal{B}, \mathcal{A})^{1/q}\|X\|_p\|Y\|_q. \qquad (b)$$

Remark 1.5 (a) was proved by Delyon (1990), (b) is due to Peligrad (1983) and implies Ibragimov's covariance inequality (cf. also Bradley and Bryc (1985), Theorem 1.1).

Proof Since (X, Y) is $\mathcal{A} \otimes \mathcal{B}$-measurable, by the polar decomposition of $\mathbb{P}_{\mathcal{A} \otimes \mathcal{B}} - \mathbb{P}_{\mathcal{A}} \otimes \mathbb{P}_{\mathcal{B}}$, we have:

$$|\mathrm{Cov}(X, Y)| \leq \int_{\Omega \times \Omega} |XY| d|\mathbb{P}_{\mathcal{A} \otimes \mathcal{B}} - \mathbb{P}_{\mathcal{A}} \otimes \mathbb{P}_{\mathcal{B}}|.$$

Let $\mu = |\mathbb{P}_{\mathcal{A} \otimes \mathcal{B}} - \mathbb{P}_{\mathcal{A}} \otimes \mathbb{P}_{\mathcal{B}}|$. By Hölder's inequality,

$$|\mathrm{Cov}(X, Y)| \le \left(\int_{\Omega \times \Omega} |X(\omega)|^p d\mu(\omega, \omega') \right)^{1/p} \left(\int_{\Omega \times \Omega} |Y(\omega')|^q d\mu(\omega, \omega') \right)^{1/q}.$$

$$(1.61)$$

Let μ_A denote the first margin of μ and μ_B the second one. Then

$$\int_{\Omega \times \Omega} |X(\omega)|^p d\mu(\omega, \omega') = \int_\Omega |X|^p d\mu_A \text{ and } \int_{\Omega \times \Omega} |Y(\omega')|^q d\mu(\omega, \omega') = \int_\Omega |Y|^q d\mu_B.$$

Hence, to prove Theorem 1.4(a), it is enough to prove that $\mu_A = 2d_A P_A$ and $\mu_B = 2d_B P_B$, for random variables d_A and d_B with the prescribed properties.

Starting from (1.58), one can prove that

$$\mu_A(A) = \sup \left\{ \sum_{i \in I} \sum_{j \in J} |\mathbb{P}(A_i \cap B_j) - \mathbb{P}(A_i)\mathbb{P}(B_j)| \right\}, \qquad (1.62)$$

the maximum being taken over all finite partitions $(A_i)_{i \in I}$ and $(B_j)_{j \in J}$ of Ω with the sets A_i in $\mathcal{A}$ and the sets B_j in $\mathcal{B}$. Therefore, for any A in $\mathcal{A}$,

$$\mu_A(A) \le \sup \left\{ \sum_{i \in I} \sum_{j \in J} (\mathbb{P}(A_i \cap B_j) + \mathbb{P}(A_i)\mathbb{P}(B_j)) \right\} \le 2\mathbb{P}(A).$$

Consequently, by the Radon–Nikodym theorem, μ_A is absolutely continuous with respect to the restriction of $\mathbb{P}$ to $\mathcal{A}$, from which it follows that $\mu_A = 2d_A\mathbb{P}$, for some nonnegative $\mathcal{A}$-measurable random variable d_A satisfying $d_A \le 1$. Finally

$$\mathbb{E}(d_A) = \int_\Omega d\mu_A = 2\beta(\mathcal{A}, \mathcal{B}),$$

which completes the proof of (a).

To prove (b), it suffices to note that $\mu_A(A) \le 2\varphi(\mathcal{A}, \mathcal{B})\mathbb{P}(A)$ by (1.59). Consequently $d_A \le \varphi(\mathcal{A}, \mathcal{B})$ a.s., which implies Theorem 1.4(b). ∎

Starting from Delyon's covariance inequality, we now give new upper bounds on the variance of partial sums of functionals. These bounds are due to Viennet (1997).

Corollary 1.4 *Let $(X_i)_{i \in \mathbb{N}}$ be a strictly stationary sequence of random variables with values in some Polish space $\mathcal{X}$. Set $\beta_i = \beta(\sigma(X_0), \sigma(X_i))$. For any numerical function g, let*

$$S_n(g) = g(X_1) + \cdots + g(X_n).$$

Denote by P the law of X_0. Then there exists a sequence $(b_i)_{i \in \mathbb{Z}}$ of measurable functions from $\mathcal{X}$ into $[0, 1]$ satisfying

$$\int_{\mathcal{X}} b_i dP = \beta_i,$$

and such that, for any function g in $L^2(P)$,

$$\mathrm{Var}\, S_n(g) \le n \int_{\mathcal{X}} (1 + 4b_1 + \cdots + 4b_{n-1})g^2 dP. \tag{a}$$

Consequently, if $B = 1 + 4\sum_{i>0} b_i$, then

$$\mathrm{Var}\, S_n(g) \le n \int_{\mathcal{X}} Bg^2 dP. \tag{b}$$

Remark 1.6 Starting from (b), one can obtain the bounds of Corollary 1.1 with the β-mixing coefficients instead of the α-mixing coefficients. Indeed, for any positive i,

$$\int_{\mathcal{X}} b_i g^2 dP = \iint_{\mathcal{X} \times [0,1]} \mathrm{1\!I}_{t \le b_i(x)} g^2(x) P \otimes \lambda(dx, dt),$$

where λ denotes the Lebesgue measure on $[0, 1]$. Let $b(t, x) = \mathrm{1\!I}_{t \le b_i(x)}$ and $h(t, x) = g^2(x)$. By (1.11b) (see Lemma 1.2, Chap. 2, for a proof of this fact),

$$\iint_{\mathcal{X} \times [0,1]} \mathrm{1\!I}_{t \le b_i(x)} g^2(x) P \otimes \lambda(dx, dt) \le \int_0^1 Q_b(u) Q_h(u) du \le \int_0^{\beta_i} Q_{g(X_0)}^2(u) du$$

(recall that $Q_h = Q_{g(X_0)}^2$). It follows that

$$\int_{\mathcal{X}} Bg^2 dP \le\le 4 \int_0^1 \beta^{-1}(u) Q_g^2(u) du. \tag{1.63}$$

Proof of Corollary 1.4 From the stationarity

$$\mathrm{Var}\, S_n(g) - n\mathrm{Var}\, g(X_0) \le 2n \sum_{i=1}^{n-1} |\mathrm{Cov}(g(X_0), g(X_i))|.$$

We now apply Theorem 1.4(a) with $p = q = 2$. There exist random variables $B_{0,i}$ and $B_{i,0}$ with values in $[0, 1]$ and with mean value β_i, measurable respectively for $\sigma(X_0)$ and $\sigma(X_i)$, such that

$$|\mathrm{Cov}(g(X_0), g(X_i))| \le 2\sqrt{\mathbb{E}(B_{0,i} g^2(X_0)) \mathbb{E}(B_{i,0} g^2(X_i))}$$
$$\le \mathbb{E}(B_{0,i} g^2(X_0)) + \mathbb{E}(B_{i,0} g^2(X_i)).$$

Now $B_{0,i} = b_{0,i}(X_0)$ and $B_{i,0} = b_{i,0}(X_i)$ and therefore, since X_0 and X_i have the common marginal law P,

$$|\mathrm{Cov}(g(X_0), g(X_i))| \le \int_{\mathcal{X}} (b_{i,0} + b_{0,i}) g^2 dP.$$

Setting $b_i = (b_{i,0} + b_{0,i})/2$, we then get (a). (b) follows immediately. ■

Corollary 1.4 yields better results for density estimation than Corollary 1.1. For example, we can relax the summability condition on the coefficients in Theorem 1.4(b), as shown by the result below, which is a particular case of the results of Viennet (1997) on L^p risks of linear estimators of the density. We refer to Dedecker and Prieur (2005) for extensions of this result to non-absolutely regular sequences.

Corollary 1.5 *Let $(X_i)_{i\in\mathbb{N}}$ be a strictly stationary sequence of random variables with values in $\mathbb{R}^d$, satisfying the assumptions of Theorem 1.4 Then, for the projection estimator of the density defined by (1.39),*

$$n \int_{\mathbb{R}^d} w(x)\text{Var }\hat{f}_n(x)dx \le (1 + 4\sum_{i=1}^{n-1}\beta_i)\sup_{x\in\mathbb{R}^d}\left(w^2(x)\sum_{j=1}^{m}e_j^2(x)\right). \qquad (b)$$

Proof By (1.41) and Corollary 1.4,

$$n \int_{\mathbb{R}^d} w(x)\text{Var }\hat{f}_n(x)dx \le \sum_{j=1}^{m}\int_{\mathbb{R}^d}\left(1 + 4\sum_{i=1}^{n-1}b_i(x)\right)f(x)w^2(x)e_j^2(x)dx,$$

with $b_i \ge 0$ and $\int_{\mathbb{R}^d} b_i(x)f(x)dx \le \beta_i$. Hence

$$n \int_{\mathbb{R}^d} w(x)\text{Var }\hat{f}_n(x)dx \le$$

$$\sup_{x\in\mathbb{R}^d}\left(w^2(x)\sum_{j=1}^{m}e_j^2(x)\right)\int_{\mathbb{R}^d}\left(1 + 4\sum_{i=1}^{n-1}b_i(x)\right)f(x)dx, \qquad (1.64)$$

which completes the proof.

1.7 Other Covariance Inequalities Under Strong Mixing *

In this section, we give sharper bounds on the covariance under strong mixing conditions. Recall that Theorem 1.1 gives upper bounds involving the quantile function of $|X|$. In this section, in order to get sharper bounds, we will use another approach. Let $\mathcal{L}_\alpha(F, G)$ denote the class of random vectors on $\mathbb{R}^2$ with given marginal distribution functions F and G, satisfying the mixing constraint $\alpha(X, Y) \le \alpha$. In the case $\alpha = 1/2$ (no mixing constraint), Fréchet (1951, 1957) and Bass (1955) proved that, for continuous distribution functions F and G,

$$\text{Cov}(X, Y) \le \int_0^1 F^{-1}(u)G^{-1}(u)du - \int_0^1 F^{-1}(u)du\int_0^1 G^{-1}(u)du, \qquad (1.65)$$

and that equality holds when $F(X) = G(Y)$ in the continuous case. In a similar way

$$\text{Cov}(X, Y) \geq \int_0^1 F^{-1}(u)G^{-1}(1 - u)du - \int_0^1 F^{-1}(u)du \int_0^1 G^{-1}(u)du, \quad (1.66)$$

and equality holds when $F(X) = 1 - G(Y)$ in the continuous case. Since the infimum of $\text{Cov}(X, Y)$ over the class $\mathcal{L}_\alpha(F, G)$ is nonpositive, the lower bound in (1.66) is nonpositive. It follows from this remark and (1.65) that

$$\text{Cov}(X, Y) \leq \int_0^1 F^{-1}(u)(G^{-1}(u) - G^{-1}(1 - u))du.$$

Next, using the change of variables $t = 1 - u$ in the integral on the right-hand side, we get that

$$\text{Cov}(X, Y) \leq \int_0^1 F^{-1}(1 - u)(G^{-1}(1 - u) - G^{-1}(u))du.$$

Define now the dispersion function D_F of F by

$$D_F(u) = F^{-1}(1 - u) - F^{-1}(u). \tag{1.67}$$

Using both the above two upper bounds on $\text{Cov}(X, Y)$, we then get that

$$\text{Cov}(X, Y) \leq \int_0^{1/2} D_F(u)D_G(u)du. \tag{1.68}$$

This upper bound is slightly suboptimal. Theorem 1.5 below gives an upper bound on the covariance involving $D_F D_G$ with a multiplicative factor $(1 - 2u)$, providing a better upper bound in the case $\alpha = 1/2$.

Theorem 1.5 *Set* $x_\alpha = (1 - \sqrt{1 - 2\alpha})/2$. *Let* (X, Y) *be an element of* $\mathcal{L}_\alpha(F, G)$. *Then*

$$|\text{Cov}(X, Y)| \leq \int_0^{x_\alpha} (1 - 2u)D_F(u)D_G(u)du. \tag{a}$$

If $(X_i)_{i \in \mathbb{Z}}$ *is a strictly stationary sequence of random variables with distribution function* F, *then*

$$|\text{Var } S_n - n\text{Var } X_0| \leq 2n \int_0^{1/2} (\alpha^{-1}(2u(1 - u)) - 1)_+ D_F^2(u)du. \tag{b}$$

Remark 1.7 Exercise 8 at the end of this chapter is devoted to a comparison between Theorems 1.1(a) and 1.5(a).

Proof The main step is the proof of (a). Without loss of generality we may assume that 0 is a median for the distributions of X and Y. For any real-valued random variable Z, let $Z^+ = \max(0, Z)$ and $Z^- = \max(0, -Z)$. Then

$$\text{Cov}(X, Y) = \text{Cov}(X^+, Y^+) + \text{Cov}(X^-, Y^-) - \text{Cov}(X^+, Y^-) - \text{Cov}(X^-, Y^+).$$

We now bound the four terms on the right-hand side. Let $H_X(x) = \mathbb{P}(X > x)$ and $H_Y(y) = \mathbb{P}(Y > y)$. From the Hoeffding identity we have:

$$\text{Cov}(X^+, Y^+) = \int_0^\infty \int_0^\infty \Big(\mathbb{P}(X > x, Y > y) - H_X(x) H_Y(y) \Big) dx\, dy.$$

Now

$$\mathbb{P}(X > x, Y > y) - H_X(x) H_Y(y) \leq \inf(H_X(x), H_Y(y)) - H_X(x) H_Y(y).$$

Let R be the increasing function defined on $[0, 1/2]$ by $R(t) = t - t^2$. Applying the elementary fact that $\inf(a, b) - ab \leq \inf(R(a), R(b))$ for any reals a and b in $[0, 1/2]$ and the strong mixing condition, we get that

$$\mathbb{P}(X > x, Y > y) - H_X(x) H_Y(y) \leq \inf(R(H_X(x)), R(H_Y(y)), \alpha/2)$$

for any positive x and y. It follows that

$$\text{Cov}(X^+, Y^+) \leq \int_0^\infty \int_0^\infty \inf(R(H_X(x)), R(H_Y(y)), \alpha/2) dx\, dy.$$

Let V be a random variable with uniform distribution over $[0, 1]$. Set

$$Z = F^{-1}(1 - R^{-1}(V)) \mathbb{1}_{V < \alpha/2} \text{ and } T = G^{-1}(1 - R^{-1}(V)) \mathbb{1}_{V < \alpha/2}.$$

Then, for any positive reals x and y,

$$\inf(R(H_X(x)), R(H_Y(y)), \alpha/2) = \mathbb{P}(Z > x, T > y).$$

Hence

$$\text{Cov}(X^+, Y^+) \leq \mathbb{E}(ZT) = \int_0^{\alpha/2} F^{-1}(1 - R^{-1}(v)) G^{-1}(1 - R^{-1}(v)) dv.$$

In the same way

$$\text{Cov}(X^-, Y^-) \leq \int_0^{\alpha/2} F^{-1}(R^{-1}(v)) G^{-1}(R^{-1}(v)) dv.$$

Now

$$-\text{Cov}(X^+, Y^-) = \int_0^\infty \int_0^\infty (H_X(x)G(-y) - \mathbb{P}(X > x, Y < -y))dxdy$$

$$\leq \int_0^\infty \int_0^\infty \inf(H_X(x)G(-y), \alpha/2)dxdy$$

$$\leq \frac{1}{2} \int_0^\infty \int_0^\infty \inf(H_X(x), G(-y), \alpha)dxdy.$$

Therefrom, proceeding as in the proof of Theorem 1.1(a),

$$-\text{Cov}(X^+, Y^-) \leq \frac{1}{2} \int_0^\alpha F^{-1}(1 - v)(-G^{-1}(v))dv$$

$$= \int_0^{\alpha/2} F^{-1}(1 - 2v)(-G^{-1}(2v))dv.$$

Now, from the convexity of R^{-1} on $[0, 1/4]$, $2v \geq R^{-1}(v)$. Since $v \to -F^{-1}(1 - v)G^{-1}(v)$ is nonincreasing on $[0, 1/2]$, we deduce from the above inequality that

$$-\text{Cov}(X^+, Y^-) \leq -\int_0^{\alpha/2} F^{-1}(1 - R^{-1}(v))G^{-1}(R^{-1}(v))dv.$$

Interchanging X and Y, we get a similar upper bound for $-\text{Cov}(X^-, Y^+)$, and, collecting the four upper bounds above, we then get that

$$\text{Cov}(X, Y) \leq \int_0^{\alpha/2} D_F(R^{-1}(v))D_G(R^{-1}(v))dv. \tag{1.69}$$

Since the dispersion function associated to the distribution function of $-X$ is also equal to D_F almost everywhere, the above upper bound still holds true for $\text{Cov}(-X, Y)$. Now Theorem 1.5(a) follows via the change of variable $u = R^{-1}(v)$.

We now prove (b). Assume that the random variables X_i have the common marginal distribution function F. With the notations of Sect. 1.4, Inequality (1.69) yields

$$|\text{Var } S_n - n\text{Var } X_0| \leq 2n \int_0^{\alpha_1/2} (\alpha^{-1}(2v) - 1)D_F^2(R^{-1}(v))dv.$$

Using again the change of variable $u = R^{-1}(v)$ in the above integral, we then get Theorem 1.5(b).

Exercises

(1) Let U be a random variable with uniform law over $[0, 1]$ and F be the distribution function of some real-valued random variable.

(a) Prove that $X = F^{-1}(U)$ has the distribution function F.

(b) Prove that, if F is continuous everywhere, then $F(X)$ has the uniform law over $[0, 1]$ and $F(X) = U$ almost surely.

(c) Let F be any distribution function (jumps are allowed) and δ be a random variable with uniform law over $[0, 1]$, independent of X. Prove that

$$V = F(X - 0) + \delta(F(X) - F(X - 0))$$

has the uniform law over $[0, 1]$, and that, almost surely $F^{-1}(V) = X$. Hint: prove that $X \geq F^{-1}(V)$, and next use the fact that X and $F^{-1}(V)$ have the same law.

(2) Let μ be a law on $\mathbb{R}^2$ and let X be a random variable with distribution the first marginal law of μ. Let δ be a random variable with uniform law over $[0, 1]$, independent of X. Construct a function f such that $(X, f(X, \delta))$ has law μ. Hint: if $Z = (T, W)$ has law μ, consider the inverse of the distribution function of W conditionally to T.

(3) Let F and G be distribution functions of nonnegative real-valued variables, and (X, Y) be a random vector with marginal distribution functions F and G.

(a) Prove that

$$\mathbb{E}(XY) \leq \int_0^1 F^{-1}(u)G^{-1}(u)du. \tag{1}$$

Suppose now that equality holds in (1). Let U be a random variable with uniform distribution over $[0, 1]$. Prove that $(F^{-1}(U), G^{-1}(U))$ and (X, Y) are equally distributed. Hint: consider the bivariate distribution function of (X, Y).

(b) Let δ be a random variable with uniform law over $[0, 1]$, independent of (X, Y). Prove that, if equality holds in (1), then one can construct a random variable $V = f(X, Y, \delta)$ with uniform law over $[0, 1]$, such that $(X, Y) = (F^{-1}(V), G^{-1}(V))$ a.s.

(4) Let X be a real-valued random variable and let Q be the quantile function of $|X|$. Let δ be a random variable with uniform law over $[0, 1]$, independent of X, and let $\mathcal{L}(\alpha)$ be the class of nonnegative integer random variables A on $(\Omega, \mathcal{T}, \mathbb{P})$, such that $\mathbb{P}(A > x) = \alpha(x)$. Prove that

$$\int_0^1 \alpha^{-1}(u)Q^2(u)du = \sup_{A \in \mathcal{L}(\alpha)} \mathbb{E}(AX^2). \tag{2}$$

(5) Throughout this exercise, the strong mixing coefficients α_n are as defined in (1.20). Let $(X_i)_{i \in \mathbb{Z}}$ be a strictly stationary sequence of real-valued random variables with law P and distribution function F. Let Z_n be defined by (1.37). We are interested

in the variance of $Z_n(I)$ for an interval I. For any Borel set A, set

$$I_n(A) = \sup\left\{\sum_{i=1}^{k} \text{Var } Z_n(A_i) : \{A_1, ..., A_k\} \text{ finite partition of } A\right\}.$$

(a) Prove that I_n is a nondecreasing and nonnegative function.

(b) Prove that, for any Borel sets A and B with $A \cap B = \emptyset$, $I_n(A \cup B) \geq I_n(A) + I_n(B)$.

(c) Prove that

$$I_n(A) \leq \sup_{\|f\|_\infty = 1} \text{Var } Z_n(f \, 1\!\!1_A). \tag{3}$$

Deduce from (3) that $I_n(\mathbb{R}) \leq 1 + 4\sum_{i=1}^{n-1} \alpha_i$.

(d) Prove that there exists some distribution function G_n such that

$$\text{Var } Z_n(]s, t]) \leq (G_n(t) - G_n(s))(1 + 4\sum_{i=1}^{n-1} \alpha_i) \tag{4}$$

for any (s, t) with $s \leq t$. Compare (4) with Corollary 1.1.

(6) Let F and G be distribution functions of nonnegative and integrable random variables and X and Y be random variables with respective distribution functions F and G. Let Φ be the set of convex functions defined in (1.26).

(a) Suppose that F and G are continuous one to one maps from $\mathbb{R}^+$ to $[0, 1[$. Prove that

$$\int_0^1 F^{-1}(u)G^{-1}(u)du = \inf_{\phi \in \Phi} \mathbb{E}(\phi^*(X) + \phi(Y)). \tag{5}$$

Hint: define ϕ by $\phi'(G^{-1}) = F^{-1}$.

(b) Does (5) hold in the general case?

(c) Let Z be a nonnegative random variable with distribution function H. Suppose that, for any ϕ in Φ, if $\phi(Y)$ is integrable, then $\phi(Z)$ is integrable. Prove that under the assumption of (a),

$$\int_0^1 F^{-1}(u)G^{-1}(u)du < \infty \implies \int_0^1 F^{-1}(u)H^{-1}(u)du < \infty.$$

(7) Let X and Y be complex-valued integrable random variables such that $|XY|$ is integrable and let $\alpha = \alpha(\sigma(X), \sigma(Y))$. Let $\mathcal{R}X$ and $\mathcal{I}X$ denote respectively the real and the imaginary part of X.

(a) Prove that $Q_{\mathcal{R}X} \leq Q_{|X|}$ and $Q_{\mathcal{I}X} \leq Q_{|X|}$.

(b) Suppose that $\mathbb{E}(XY) - \mathbb{E}(X)\mathbb{E}(Y) = \rho \geq 0$. Apply Theorem 1.1 to the real and the imaginary parts of X and Y to prove that

$$|\mathbb{E}(XY) - \mathbb{E}(X)\mathbb{E}(Y)| \le 4 \int_0^\alpha Q_{|X|}(u) Q_{|Y|}(u) du. \tag{6}$$

(c) **The general case.** Suppose that $\mathbb{E}(XY) - \mathbb{E}(X)\mathbb{E}(Y) = \rho e^{i\theta}$ for some $\rho > 0$ and some θ in $\mathbb{R}$. Apply (b) to X and $e^{-i\theta}Y$ to prove that Inequality (6) still holds true.

(8) Let X and Y be two random variables, with respective distribution functions F and G, satisfying the assumptions of Theorem 1.1 or Theorem 1.5.

(a) Prove that, for any (x, y) in $\mathbb{R}^2$,

$$|\mathbb{P}(X > x, Y > y) - \mathbb{P}(X > x)\mathbb{P}(y > y)| \le \inf(F(x), G(x), 1 - F(x), 1 - G(x), \alpha/2).$$

(b) With the notations of Theorem 1.5, infer from the above inequality that

$$|\mathrm{Cov}(X, Y)| \le \int_0^{\alpha/2} D_F(u) D_G(u) du. \tag{7}$$

(c) Noticing that the upper bound in (1.69) is equal to the upper bound in Theorem 1.5(a), prove that Theorem 1.5(a) is sharper than (7). Hint: prove that $R^{-1}(v) \ge v$.

(d) Symmetric case. Assume here that X and Y have symmetric laws. Prove then that $D_F(u) = 2Q_X(2u)$ and $D_G(u) = 2Q_Y(2u)$ almost everywhere. Infer that

$$\int_0^{\alpha/2} D_F(u) D_G(u) du = 2 \int_0^\alpha Q_X(u) Q_Y(u) du.$$

(e) General case. For a real-valued random variable Z, define Ψ_Z by $\Psi_Z(x) = \mathbb{P}(Z > x)$ for $x \ge 0$ and $\Psi_Z(x) = \mathbb{P}(Z < x)$ for $x < 0$. Go inside the paper of Rio (1993, pp. 593–594) to prove that

$$2 \int_0^\alpha Q_X(u) Q_Y(u) du \ge \iint_{\mathbb{R}^2} \inf(\Psi_X(x), \Psi_Y(y), \alpha/2) dx dy.$$

Infer that (7) is sharper than Theorem 1.1(a) in the general case.

Chapter 2
Algebraic Moments, Elementary Exponential Inequalities

2.1 Introduction

In this chapter, we start by giving upper bounds for algebraic moments of partial sums from a strongly mixing sequence. These inequalities are similar to Rosenthal's inequalities (1970) concerning moments of sums of independent random variables. They may be applied to provide estimates of deviation probabilities of partial sums from their mean value, which are more efficient than the results derived from the Marcinkiewicz–Zygmund type moment inequalities given in Ibragimov (1962) or Billingsley (1968) for uniformly mixing sequences, or in Yokoyama (1980) for strongly mixing sequences, in particular for partial sums with a small variance. For example, Rosenthal type inequalities may be used to obtain precise upper bounds for integrated L^p-risks of kernel density estimators. They provide the exact rates of convergence, in contrast to Marcinkiewicz–Zygmund type moment inequalities, as shown first by Bretagnolle and Huber (1979) in the independent case.

In Sects. 2.2 and 2.3, we follow the approach of Doukhan and Portal (1983), for algebraic moments in the strong mixing case. In Sect. 2.4 we give a second method, which provides explicit constants in inequalities for the algebraic moments of order $2p$. Applying then the Markov inequality to S_n^{2p}, and minimizing the so obtained deviation bound with respect to p, we then get exponential Hoeffding type inequalities in the uniform mixing case. We also apply this method to obtain upper bounds for non-algebraic moments in Sect. 2.5.

2.2 An Upper Bound for the Fourth Moment of Sums

In this section, we adapt the method introduced in Billingsley (1968, Sect. 22) to bound the moment of order 4 of a sum of random variables satisfying a uniform mixing condition in the context of strongly mixing sequences. We start by introducing some notation that we shall use throughout the sequel.

© Springer-Verlag GmbH Germany 2017
E. Rio, *Asymptotic Theory of Weakly Dependent Random Processes*,
Probability Theory and Stochastic Modelling 80,
DOI 10.1007/978-3-662-54323-8_2

Notation 2.1 Let $(X_i)_{i \in \mathbb{Z}}$ be a sequence of real-valued random variables. Set $\mathcal{F}_k = \sigma(X_i : i \leq k)$ and $\mathcal{G}_l = \sigma(X_i : i \geq l)$. By convention, if the sequence $(X_i)_{i \in T}$ is defined on a subset T of $\mathbb{Z}$, we set $X_i = 0$ for i in $\mathbb{Z} \setminus T$.

Throughout Sects. 2.2 and 2.3, the strong mixing coefficients $(\alpha_n)_{n \geq 0}$ of $(X_i)_{i \in \mathbb{Z}}$ are defined, as in Rosenblatt (1956), by

$$\alpha_0 = 1/2 \text{ and } \alpha_n = \sup_{k \in \mathbb{Z}} \alpha(\mathcal{F}_k, \mathcal{G}_{k+n}) \text{ for any } n > 0. \qquad (2.1)$$

Starting from Theorem 1.1(a), we now give an upper bound for the fourth moment of the partial sums for nonstationary sequences.

Theorem 2.1 *Let $(X_i)_{i \in \mathbb{N}}$ be a sequence of centered real-valued random variables with finite fourth moments. Let $Q_k = Q_{|X_k|}$ and set*

$$M_{4,\alpha,n}(Q_k) = \sum_{k=1}^{n} \int_0^1 [\alpha^{-1}(u) \wedge n]^3 Q_k^4(u) du.$$

Then

$$\mathbb{E}(S_n^4) \leq 3 \left(\sum_{i=1}^{n} \sum_{j=1}^{n} |\mathbb{E}(X_i X_j)| \right)^2 + 48 \sum_{k=1}^{n} M_{4,\alpha,n}(Q_k).$$

Proof For $i \notin [1, n]$, let us replace the initial random variables X_i by the null random variable. With this convention

$$S_n^4 = 24 \sum_{i<j<k<l} X_i X_j X_k X_l + 12 \sum_{\substack{j<k \\ i \notin \{j,k\}}} X_i^2 X_j X_k + 6 \sum_{i<j} X_i^2 X_j^2 + 4 \sum_{i \neq j} X_i^3 X_j + \sum_i X_i^4. \qquad (2.2)$$

It follows that

$$\mathbb{E}(S_n^4) \leq 3 \sum_{i \leq j \leq k \leq l} |\mathbb{E}(X_i X_j X_k X_l)|(1 + \mathrm{I\!I}_{i<j})(1 + \mathrm{I\!I}_{j<k})(1 + \mathrm{I\!I}_{k<l}). \qquad (2.3)$$

We now apply Theorem 1.1(a) to the product $X_i X_j X_k X_l$ at the maximal spacing. So, let $m = \sup(j - i, k - j, l - k)$. If $m = k - j > 0$, then Theorem 1.1(a) applied to $X = X_i X_j$ and $Y = X_k X_l$ yields

$$|\mathbb{E}(X_i X_j X_k X_l)| \leq |\mathbb{E}(X_i X_j) \mathbb{E}(X_k X_l)| + 2 \int_0^{\alpha_m} Q_{X_i X_j}(u) Q_{X_k X_l}(u) du. \qquad (2.4)$$

If $m = j - i$ and $k - j < m$, Theorem 1.1(a) applied to $X = X_i$ and $Y = X_j X_k X_l$ yields

$$|\mathbb{E}(X_i X_j X_k X_l)| \leq 2 \int_0^{\alpha_m} Q_{X_i}(u) Q_{X_j X_k X_l}(u) du. \qquad (2.5)$$

The case $m = l - k$ and $\sup(k - j, j - i) < m$ can be treated in the same way and gives the same inequality. To complete the proof, we will need the technical lemma below, due to Bass (1955) in the case $p = 2$.

Lemma 2.1 *Let $Z_1, \ldots Z_p$ be nonnegative random variables. Then*

$$\mathbb{E}(Z_1 \ldots Z_p) \le \int_0^1 Q_{Z_1}(u) \ldots Q_{Z_p}(u)du. \qquad (a)$$

Furthermore,

$$\int_0^1 Q_{Z_1 Z_2}(u) Q_{Z_3}(u) \ldots Q_{Z_p}(u)du \le \int_0^1 Q_{Z_1}(u) Q_{Z_2}(u) \ldots Q_{Z_p}(u)du \qquad (b)$$

and

$$\int_0^1 Q_{Z_1 + Z_2}(u) Q_{Z_3}(u) \ldots Q_{Z_p}(u)du \le \int_0^1 (Q_{Z_1}(u) + Q_{Z_2}(u)) Q_{Z_3}(u) \ldots Q_{Z_p}(u)du. \qquad (c)$$

Proof of Lemma 2.1 We first prove (a). By Fubini's Theorem,

$$\mathbb{E}(Z_1 \ldots Z_p) = \int_{\mathbb{R}^p} \mathbb{P}(Z_1 > z_1, \ldots, Z_p > z_p)dz_1 \ldots dz_p$$

$$\le \int_{\mathbb{R}^p} \inf_{i \in [1,p]} \mathbb{P}(Z_i > z_i)dz_1 \ldots dz_p. \qquad (2.6)$$

Now

$$\inf_{i \in [1,p]} \mathbb{P}(Z_i > z_i) = \int_0^1 \mathbb{1}_{z_1 < Q_{Z_1}(u)} \ldots \mathbb{1}_{z_p < Q_{Z_p}(u)}du. \qquad (2.7)$$

Plugging (2.7) into (2.6) and again applying Fubini's theorem, we then get (a).

Let us now prove (b). Let U be a random variable with the uniform distribution over $[0, 1]$. For any nonnegative random variable Z, $Q_Z(U)$ has the distribution of Z. Now (cf. Exercise 1, Chap. 1), if $H(t) = \mathbb{P}(Z_1 Z_2 > t)$, then, for any random variable δ with uniform distribution over $[0, 1]$ independent of (Z_1, Z_2),

$$W = 1 - V = H(Z_1 Z_2 - 0) + \delta(H(Z_1 Z_2) - H(Z_1 Z_2 - 0))$$

has the uniform law. Let $(T_1, T_2, \cdots, T_p) = (Z_1, Z_2, Q_{Z_3}(W), \ldots, Q_{Z_p}(W))$. Then the random variable $(T_1 T_2, T_3, \ldots, T_p)$ has the same law as $(Q_{Z_1 Z_2}(U), Q_{Z_3}(U), \ldots, Q_{Z_p}(U))$. Hence, by Lemma 2.1(a),

$$\int_0^1 Q_{Z_1 Z_2}(u) Q_{Z_3}(u) \ldots Q_{Z_p}(u)du \le \int_0^1 Q_{Z_1}(u) Q_{Z_2}(u) \ldots Q_{Z_p}(u)du,$$

which completes the proof of (b). The proof of (c), being similar, is omitted. ■

We now complete the proof of Theorem 2.1. Both inequalities (2.4) and (2.5) together with Lemma 2.1(b) applied repeatedly yield

$$
|\mathbb{E}(X_i X_j X_k X_l)| \leq 2 \int_0^{\alpha_m} Q_i(u) Q_j(u) Q_k(u) Q_l(u)\,du
$$
$$
+ |\mathbb{E}(X_i X_j)\mathbb{E}(X_k X_l)|\, \mathrm{II}_{k-j>\max(j-i,l-k)}, \qquad (2.8)
$$

where $m = \max(j - i, k - j, l - k) > 0$ is the maximal spacing. In the case $m = 0$, (2.8) still holds since

$$
E(X_i^4) = \int_0^1 Q_i^4(u)\,du \leq 2 \int_0^{1/2} Q_i^4(u)\,du.
$$

Now

$$
\sum_{i \leq j < k \leq l} |\mathbb{E}(X_i X_j)\mathbb{E}(X_k X_l)|(1 + \mathrm{II}_{i<j})(1 + \mathrm{II}_{k<l}) \leq \Big(\sum_{(i,j)\in[1,n]^2} |\mathbb{E}(X_i X_j)| \Big)^2.
$$

Hence, by (2.3) and (2.8),

$$
\mathbb{E}(S_n^4) - 3\Big(\sum_{i=1}^n \sum_{j=1}^n |\mathbb{E}(X_i X_j)|\Big)^2 \leq 12 \sum_{i \leq j \leq k \leq l} \int_0^{\alpha_m} (Q_i^4(u) + Q_j^4(u) + Q_k^4(u) + Q_l^4(u))\,du
$$
$$
\leq 48 \sum_{m=0}^{n-1} \sum_{k=1}^n \int_{\alpha_{m+1}}^{\alpha_m} (m+1)^3 Q_k^4(u)\,du, \qquad (2.9)
$$

with the convention that $\alpha_n = 0$ in (2.9). Hence Theorem 2.1 holds ■

Application of Theorem 2.1 to bounded random variables. Suppose that $\|X_i\|_\infty \leq 1$ for any $i > 0$. Then by Theorem 2.1 and Corollary 1.2,

$$
\mathbb{E}(S_n^4) \leq 3\Big(\sum_{i=1}^n \sum_{j=1}^n |\mathbb{E}(X_i X_j)|\Big)^2 + 144n \sum_{m=0}^{n-1} (m+1)^2 \alpha_m
$$
$$
\leq 48n^2 \Big(\sum_{m=0}^{n-1} \alpha_m\Big)^2 + 144n \sum_{m=0}^{n-1} (m+1)^2 \alpha_m. \qquad (2.10)
$$

Let us compare this result with Lemma 4, Sect. 20, in Billingsley (1968). This lemma gives, in our setting (note that Billingsley's proof can be adapted to strongly mixing sequences),

$$
\mathbb{E}(S_n^4) \leq 768n^2 \Big(\sum_{m=0}^{n-1} \sqrt{\alpha_m}\Big)^2. \qquad (2.11)
$$

For any $p > 0$, set

$$\Lambda_p(\alpha^{-1}) = \sup_{0 \leq m < n} (m+1)(\alpha_m)^{1/p}. \tag{2.12}$$

Applying (2.10), we get

$$\mathbb{E}(S_n^4) \leq (8\pi^2 + 144)(n\Lambda_2(\alpha^{-1}))^2 \leq 223n^2(\Lambda_2(\alpha^{-1}))^2. \tag{2.13}$$

Since $(\alpha_m)_{m \geq 0}$ is nonincreasing,

$$\Lambda_2(\alpha^{-1}) \leq \sum_{m=0}^{n-1} \sqrt{\alpha_m}, \tag{2.14}$$

which shows that (2.13) implies (2.11). Now, if the strong mixing coefficients α_m satisfy $\alpha_m = O(m^{-2})$, then (2.13) ensures that $\mathbb{E}(S_n^4) = O(n^2)$. In that case (2.11) leads to a logarithmic loss.   ∎

2.3   Even Algebraic Moments

In this section, we extend Theorem 2.1 to moments of order $2p$ with $p > 2$ an integer. Our main result is the following.

Theorem 2.2 *Let $p > 0$ be an integer and $(X_i)_{i \in \mathbb{N}}$ be a sequence of centered real-valued random variables with finite moments of order $2p$. Set $Q_k = Q_{X_k}$. Then there exist positive constants a_p and b_p such that*

$$\mathbb{E}\left(S_n^{2p}\right) \leq a_p \left(\int_0^1 \sum_{k=1}^n [\alpha^{-1}(u) \wedge n] Q_k^2(u) du\right)^p$$

$$+ b_p \sum_{k=1}^n \int_0^1 [\alpha^{-1}(u) \wedge n]^{2p-1} Q_k^{2p}(u) du.$$

Remark 2.1 Recall that $Q_k(U)$ and $|X_k|$ have the same law. The weighted moments on the right-hand side of the above inequality play the same role as the usual moments in the independent case. We refer to Annex C for more comparisons between these quantities and the usual moments.

Doukhan and Portal (1983) give recursive relations which allow us to bound a_p and b_p by induction on p. These upper bounds can be used to derive exponential inequalities for geometrically strongly mixing sequences or random fields (cf. Doukhan et al. (1984) or Doukhan 1994). For nonalgebraic moments, one can derive moment inequalities from the algebraic case via interpolation inequalities (see Utev (1985) or Doukhan 1994). Nevertheless, interpolation inequalities lead to suboptimal mixing conditions. In Chap. 6, we will give another way to prove moment inequalities, which leads to unimprovable mixing conditions.

Proof of Theorem 2.2 We follow the line of proof of Doukhan and Portal (1983); cf. also Doukhan (1994). For any positive integer q, let

$$A_q(n) = \sum_{1 \le i_1 \le \cdots \le i_q \le n} |\mathbb{E}(X_{i_1} \ldots X_{i_q})|. \tag{2.15}$$

It is easy to check that

$$\mathbb{E}(S_n^{2p}) \le (2p)! A_{2p}(n). \tag{2.16}$$

Theorem 2.2 then follows from similar upper bounds on $A_q(n)$. We will bound these quantities by induction on q via Lemma 2.2 below.

Lemma 2.2 *Suppose that the random variables* $X_1, \ldots X_n$ *are centered and with finite absolute moments of order* q. *Then*

$$A_q(n) \le \sum_{r=1}^{q-1} A_r(n) A_{q-r}(n) + 2 \sum_{k=1}^{n} \int_0^1 [\alpha^{-1}(u) \wedge n]^{q-1} Q_k^q(u) du.$$

Proof As in the proof of Theorem 2.1, we may assume that $\alpha_n = 0$. Let

$$m(i_1, \ldots, i_q) = \sup_{k \in [1, q[} (i_{k+1} - i_k)$$

and

$$j = \inf\{k \in [1, q[: i_{k+1} - i_k = m(i_1, \ldots, i_q)\}. \tag{2.17}$$

Theorem 1.1(a) applied to $X = X_{i_1} \ldots X_{i_j}$ and $Y = X_{i_{j+1}} \ldots X_{i_q}$ together with Lemma 2.1(b) ensures that

$$|\mathbb{E}(X_{i_1} \ldots X_{i_q})| \le |\mathbb{E}(X_{i_1} \ldots X_{i_j}) \mathbb{E}(X_{i_{j+1}} \ldots X_{i_q})| + 2 \int_0^{\alpha_{m(i_1,\ldots,i_q)}} Q_{i_1}(u) \ldots Q_{i_q}(u) du. \tag{2.18}$$

Summing (2.18) over $(i_1, \ldots, i_q)$ we infer that

$$A_q(n) \le \sum_{r=1}^{q-1} A_r(n) A_{q-r}(n) + 2 \sum_{i_1 \le \cdots \le i_q} \int_0^{\alpha_{m(i_1,\ldots,i_q)}} Q_{i_1}(u) \ldots Q_{i_q}(u) du. \tag{2.19}$$

Now, starting from the elementary inequality

$$Q_{i_1}(u) \ldots Q_{i_q}(u) \le q^{-1}(Q_{i_1}^q(u) + \cdots + Q_{i_q}(u)),$$

and interchanging the sum and the integral, we get that

$$\sum_{i_1 \le \cdots \le i_q} \int_0^{\alpha_{m(i_1,\ldots,i_q)}} Q_{i_1}(u) \ldots Q_{i_q}(u) du \le \frac{1}{q} \sum_{l=1}^{q} \sum_{i_l=1}^{n} \sum_{m=0}^{n-1} \int_{\alpha_{m+1}}^{\alpha_m} \chi(i_l, m) Q_{i_l}^q(u) du,$$

where $\chi(i_l, m)$ is the cardinality of the set of $(q-1)$-tuples $(i_1, .., i_{l-1}, i_{l+1}, .., i_q)$ such that

$$i_1 \le \cdots \le i_{l-1} \le i_l \le i_{l+1} \le \cdots \le i_q \quad \text{and} \quad \sup_{k \in [1,q[} (i_{k+1} - i_k) \le m.$$

Noting that $\chi(i_l, m) \le (m+1)^{q-1}$, we then get Lemma 2.2. ■

End of the proof of Theorem 2.2. Let

$$M_{q,\alpha,n} = \sum_{k=1}^{n} \int_0^1 [\alpha^{-1}(u) \wedge n]^{q-1} Q_k^q(u) du. \tag{2.20}$$

We will prove by induction on q that

$$A_q(n) \le a_q M_{2,\alpha,n}^{q/2} + b_q M_{q,\alpha,n}. \tag{$\mathcal{H}(q)$}$$

By Corollary 1.2, $\mathcal{H}(2)$ holds true with $a_2 = 2$ and $b_2 = 0$. Suppose now that $\mathcal{H}(r)$ holds for any $r \le q - 1$ Then, from Lemma 2.2 we get that

$$A_q(n) \le \sum_{r=2}^{q-2} (a_r M_{2,\alpha,n}^{r/2} + b_r M_{r,\alpha,n})(a_{q-r} M_{2,\alpha,n}^{(q-r)/2} + b_{q-r} M_{q-r,\alpha,n}) + 2M_{q,\alpha,n}.$$

Hence $\mathcal{H}(q)$ will hold true if we prove that, for any r in $[2, q-2]$,

$$(a_r M_{2,\alpha,n}^{r/2} + b_r M_{r,\alpha,n})(a_{q-r} M_{2,\alpha,n}^{(q-r)/2} + b_{q-r} M_{q-r,\alpha,n}) \le a_{q,r} M_{2,\alpha,n}^{q/2} + b_{q,r} M_{q,\alpha,n}. \tag{2.21}$$

To prove (2.21) we apply Young's inequality $qxy \le rx^{q/r} + (q - r)y^{q/(q-r)}$ to the left-hand side of (2.21). Noting that $(v + w)^s \le 2^{s-1}(v^s + w^s)$ for any $s \ge 1$, we get that (2.21) will hold true if

$$M_{r,\alpha,n}^{q/r} \le c_{q,r} (M_{2,\alpha,n}^{q/2} + M_{q,\alpha,n}). \tag{2.22}$$

Now, let

$$M_{p,\alpha,n}(Q_k) = \int_0^1 [\alpha^{-1}(u) \wedge n]^{p-1} Q_k^p(u) du.$$

By Hölder's inequality,

$$M_{r,\alpha,n}(Q_k) \le (M_{q,\alpha,n}(Q_k))^{(r-2)/(q-2)} (M_{2,\alpha,n}(Q_k))^{(q-r)/(q-2)}.$$

Therefore

$$M_{r,\alpha,n} = \sum_{k=1}^{n} M_{r,\alpha,n}(Q_k) \le \sum_{k=1}^{n} (M_{q,\alpha,n}(Q_k))^{(r-2)/(q-2)} (M_{2,\alpha,n}(Q_k))^{(q-r)/(q-2)}.$$

Hence, by Hölder's inequality applied with exponents $(q-2)/(r-2)$ and $(q-2)/(q-r)$ together with the appropriate Young's inequality,

$$M_{r,\alpha,n} \le M_{q,\alpha,n}^{(r-2)/(q-2)} M_{2,\alpha,n}^{(q-r)/(q-2)} \le c'_{r,q} (M_{q,\alpha,n}^{r/q} + M_{2,\alpha,n}^{r/2}),$$

which implies (2.22). Whence (2.21) holds, and Lemma 2.2 follows by induction on q. Both (2.16) and Lemma 2.2 then imply Theorem 2.2. ∎

Application to bounded random variables. Suppose that $\|X_i\|_\infty \le 1$ for any $i > 0$. Then

$$\mathbb{E}(S_n^{2p}) \le (2a_p + b_p)n^p (\Lambda_p(\alpha^{-1}))^p. \tag{2.23}$$

Consequently, if the strong mixing coefficients $(\alpha_m)_{m\ge0}$ satisfy $\alpha_m = O(m^{-p})$, then (2.23) implies the Marcinkiewicz–Zygmund type inequality $\mathbb{E}(S_n^{2p}) = O(n^p)$. In that case Yokoyama's inequalities (1980) are not efficient (cf. Annex C for more details). ∎

2.4 Exponential Inequalities

The constants a_p and b_p appearing in Theorem 2.2 can be bounded by explicit constants. Nevertheless, in the case of geometrically mixing sequences, it seems that it is difficult to obtain the exact dependence in p of the constants (recall that one can derive exponential inequalities from moment inequalities with explicit constants). In this section, we give a different way to obtain moment inequalities, which is more suitable for deriving exponential inequalities. Next we will derive exponential inequalities for geometrically strongly mixing inequalities from these new inequalities. We will also obtain the so-called Collomb inequalities (1984) for uniformly mixing sequences via this method. We refer to Delyon (2015) and Wintenberger (2010) for additional results.

Notation 2.2 Let $\mathcal{F}_i = \sigma(X_j : j \le i)$. We set $\mathbb{E}_i(X_k) = \mathbb{E}(X_k \mid \mathcal{F}_i)$.

The fundamental tool of this section is the equality below.

Theorem 2.3 *Let $(X_i)_{i\in\mathbb{Z}}$ be a sequence of real-valued random variables and ψ be a convex differentiable map from $\mathbb{R}$ into $\mathbb{R}^+$, with $\psi(0) = 0$, and such that the second derivative of ψ in the sense of distributions is absolutely continuous with respect to the Lebesgue measure on $\mathbb{R}$. Let ψ'' denote the density of the second derivative of ψ. Suppose that for any i in $[1, n]$ and any k in $[i, n]$,*

$$\mathbb{E}(|(\psi'(S_i) - \psi'(S_{i-1}))X_k|) < \infty. \tag{a}$$

Then

$$\mathbb{E}(\psi(S_n)) = \sum_{i=1}^{n} \int_0^1 \mathbb{E}\left(\psi''(S_{i-1} + tX_i)\left(-tX_i^2 + X_i \sum_{k=i}^{n} \mathbb{E}_i(X_k)\right)\right) dt.$$

Proof By the Taylor integral formula of order 2

$$\psi(S_n) = \sum_{i=1}^{n}(\psi(S_i) - \psi(S_{i-1}))$$

$$= \sum_{k=1}^{n} \psi'(S_{k-1})X_k + \sum_{i=1}^{n} \int_0^1 (1-t)\psi''(S_{i-1} + tX_i)X_i^2 dt.$$

Now

$$\psi'(S_{k-1}) = \sum_{i=1}^{k-1}(\psi'(S_i) - \psi'(S_{i-1})) = \sum_{i=1}^{k-1} \int_0^1 \psi''(S_{i-1} + tX_i)X_i dt.$$

Plugging this equality into the Taylor formula, we get that

$$\psi(S_n) = \sum_{i=1}^{n} \int_0^1 \psi''(S_{i-1} + tX_i)\left(-tX_i^2 + X_i \sum_{k=i}^{n} X_k\right) dt. \tag{2.24}$$

Now, taking the mean of the above equality, noticing that, under assumption (a), the random variables $(1-t)\psi''(S_{i-1} + tX_i)X_i^2$ and $\psi''(S_{i-1} + tX_i)X_iX_k$ are integrable with respect to the product measure $\lambda \otimes \mathbb{P}$ and applying Fubini's theorem, we get that

$$\mathbb{E}(\psi(S_n)) = \sum_{i=1}^{n} \int_0^1 \mathbb{E}\left(\psi''(S_{i-1} + tX_i)\left(-tX_i^2 + X_i \sum_{k=i}^{n} X_k\right)\right) dt.$$

Theorem 2.3 then follows from this equality and the fact that

$$\mathbb{E}\left(\psi''(S_{i-1} + tX_i)X_iX_k\right) = \mathbb{E}\left(\psi''(S_{i-1} + tX_i)X_i\mathbb{E}_i(X_k)\right). \quad \blacksquare$$

We now derive a Hoeffding type inequality from Theorem 2.3 (cf. Theorem B.4, Annex B, for Hoeffding's inequality for bounded and independent random variables). This inequality is an extension of the Azuma inequality (1967) for martingales to dependent sequences.

Theorem 2.4 *Let $(X_i)_{i\in\mathbb{Z}}$ be a sequence of real-valued bounded random variables and let $(m_1, m_2, \ldots, m_n)$ be an n-tuple of positive reals such that*

$$\sup_{j\in[i,n]} \left(\|X_i^2\|_\infty + 2\|X_i \sum_{k=i+1}^{j} \mathbb{E}_i(X_k)\|_\infty \right) \le m_i \ \text{ for any } i \in [1,n], \qquad (a)$$

with the convention $\sum_{k=i+1}^{i} \mathbb{E}_i(X_k) = 0$. Then, for any nonnegative integer p,

$$\mathbb{E}(S_n^{2p}) \le \frac{(2p)!}{2^p\, p!} \left(\sum_{i=1}^{n} m_i \right)^p. \qquad (b)$$

Consequently, for any positive x,

$$\mathbb{P}(|S_n| \ge x) \le \sqrt{e} \exp\left(-x^2/(2m_1 + \cdots + 2m_n) \right). \qquad (c)$$

Proof Define the functions ψ_p by $\psi_0(x) = 1$ and $\psi_p(x) = x^{2p}/(2p)!$ for $p > 0$. Set $M_i = \|X_i\|_\infty^2$. We prove (b) by induction on p. At range 0, (b) holds true for any sequence $(X_i)_{i\in\mathbb{Z}}$, since $S_n^0 = 1$. If (b) holds at range p for any sequence $(X_i)_{i\in\mathbb{Z}}$, then, applying Theorem 2.3 to $\psi = \psi_{p+1}$ and noting that $\psi_{p+1}'' = \psi_p$, we get that

$$2\mathbb{E}(\psi_{p+1}(S_n)) \le \sum_{i=1}^{n} \int_0^1 \mathbb{E}(\psi_p(S_{i-1} + tX_i))(m_i + (1-2t)M_i)dt. \qquad (2.25)$$

We now apply the induction hypothesis to the sequence $(X_l')_{l\in\mathbb{Z}}$ defined by $X_l' = X_l$ for any $1 \le l < i$, $X_i' = tX_i$ and $X_l' = 0$ for $l \notin [1,i]$. For $l < i$ and $j < i$,

$$X_l' \sum_{m=l+1}^{j} \mathbb{E}_l(X_m') = X_l \sum_{m=l+1}^{j} \mathbb{E}_l(X_m).$$

For $l < i$ and $j \ge i$,

$$X_l' \sum_{m=l+1}^{j} \mathbb{E}_l(X_m') = tX_l \sum_{m=l+1}^{i} \mathbb{E}_l(X_m) + (1-t)X_l \sum_{m=l+1}^{i-1} \mathbb{E}_l(X_m).$$

Hence the sequence $(X_l')_{l\in\mathbb{Z}}$ satisfies assumption (a) with the new sequence $(m_i')_i$ defined by $m_l' = m_l$ for $l < i$ and $m_i' = t^2 M_i$. Consequently, applying (b) to $S_i' = X_1' + \cdots + X_i'$, we get that

$$2^p\, p!\, \mathbb{E}(\psi_p(S_{i-1} + tX_i)) \le (m_1 + \cdots + m_{i-1} + t^2 M_i)^p.$$

Now $m_i + (1-2t)M_i \ge m_i - M_i \ge 0$. Hence

$$2^{p+1} p! \int_0^1 \mathbb{E}(\psi_p(S_{i-1} + tX_i))(m_i + (1 - 2t)M_i)dt$$

$$\leq \int_0^1 (m_1 + \cdots + m_{i-1} + t^2 M_i)^p (m_i + (1 - 2t)M_i)dt$$

$$\leq \int_0^1 (m_1 + \cdots + m_{i-1} + tm_i + t(1-t)M_i)^p (m_i + (1 - 2t)M_i)dt, \tag{2.26}$$

since $tm_i + t(1-t)M_i \geq t^2 M_i$. Now

$$(p+1) \int_0^1 (m_1 + \cdots + m_{i-1} + tm_i + t(1-t)M_i)^p (m_i + (1-2t)M_i)dt =$$

$$(m_1 + \cdots + m_i)^{p+1} - (m_1 + \cdots + m_{i-1})^{p+1}, \tag{2.27}$$

whence

$$2^{p+1}(p+1)! \int_0^1 \mathbb{E}(\psi_p(S_{i-1} + tX_i))(m_i + (1 - 2t)M_i)dt \leq$$

$$(m_1 + \cdots + m_i)^{p+1} - (m_1 + \cdots + m_{i-1})^{p+1}. \tag{2.28}$$

Finally, both (2.25) and (2.28) ensure that the induction hypothesis holds at range $p + 1$ for the sequence $(X_i)_{i \in \mathbb{Z}}$. Hence (b) holds true by induction on p.

In order to prove (c), we will apply the Markov inequality to S_n^{2p} for some appropriate p. Set

$$A = x^2/(2m_1 + \cdots + 2m_n) \quad \text{and} \quad p = [A + (1/2)],$$

the square brackets designating the integer part. (c) holds trivially for $A \leq 1/2$. Hence we may assume that $A \geq 1/2$. Then $p > 0$, and applying the Markov inequality to S_n^{2p}, we get that

$$\mathbb{P}(|S_n| \geq x) \leq (4A)^{-p}(2p)!/p! \tag{2.29}$$

If A belongs to $[1/2, 3/2]$, (2.29) yields

$$\mathbb{P}(|S_n| \geq x) \leq (2A)^{-1} \leq \sqrt{e} \exp(-A),$$

since $2A \geq \exp(A - 1/2)$ for A in $[1/2, 3/2]$. Next, if $A \geq 3/2$, using the fact that the sequence $(2\pi n)^{-1/2}(e/n)^n n!$ is nonincreasing, we get that $(2p)! \leq \sqrt{2}(4p/e)^p p!$, whence

$$\mathbb{P}(|S_n| \geq x) \leq \sqrt{2}(eA)^{-p} p^p.$$

Now, taking the logarithm of this inequality, we obtain

$$A + \log \mathbb{P}(|S_n| \geq x) \leq \log \sqrt{2} + f_p(A),$$

with $f_p(A) = (A - p) - p \log(A/p)$. Here $p \geq 2$ and A belongs to $[p - 1/2, p + 1/2[$. Since $f_p'(A) = (A - p)/A$ and $f_p''(A) = p/A^2$, the function f_p is convex. Consequently the maximum of f_p is attained at $A = p - 1/2$ or $A = p + 1/2$. Since f_p reaches its minimum at point p and f_p'' is decreasing, the maximum of f_p is attained for $A = p - 1/2$. Hence

$$A + \log \mathbb{P}(|S_n| \geq x) \leq \frac{\log 2 - 1}{2} + p \log\left(\frac{2p}{2p - 1}\right) \leq \frac{\log 2 - 1}{2} + 2\log(4/3),$$

since $p \geq 2$. Thus we get that

$$\mathbb{P}(|S_n| \geq x) \leq \frac{16\sqrt{2}}{9\sqrt{e}} \exp(-A) \leq \sqrt{e} \exp(-A),$$

which completes the proof of Theorem 2.4(c). ∎

We now apply Theorem 2.4 to uniformly mixing sequences, as defined below.

Definition 2.1 The uniform mixing coefficients of $(X_i)_{i \in \mathbb{Z}}$ are defined by

$$\varphi_0 = 1 \quad \text{and} \quad \varphi_n = \sup_{k \in \mathbb{Z}} \varphi(\mathcal{F}_k, \sigma(X_{k+n})) \quad \text{for any } n > 0.$$

The sequence $(X_i)_{i \in \mathbb{Z}}$ is said to be uniformly mixing if φ_n converges to 0 as n tends to ∞.

Corollary 2.1 below provides a Hoeffding type inequality for uniformly mixing sequences of bounded random variables.

Corollary 2.1 *Let* $(X_i)_{i \in \mathbb{Z}}$ *be a sequence of centered and real-valued bounded random variables. Set* $\theta_n = 1 + 4(\varphi_1 + \cdots + \varphi_{n-1})$ *and* $M_i = \|X_i\|_\infty^2$. *Then, for any positive integer* p,

$$\mathbb{E}(S_n^{2p}) \leq \frac{(2p)!}{p!} \left(\frac{\theta_n}{2}\right)^p (M_1 + \cdots + M_n)^p. \tag{a}$$

Next, for any positive x,

$$\mathbb{P}(|S_n| \geq x) \leq \sqrt{e} \exp\left(-x^2/(2\theta_n M_1 + \cdots + 2\theta_n M_n)\right). \tag{b}$$

Proof Let us apply Theorem 2.4 to the sequence $(X_i)_{i \in \mathbb{Z}}$. Since the random variables X_k are centered at expectation, by Theorem 1.4(b) and the Riesz–Fisher theorem,

$$\|\mathbb{E}_i(X_k)\|_\infty \leq 2\varphi_{k-i}\|X_k\|_\infty.$$

Hence we may apply Theorem 2.4 with

$$m_i = M_i + 4 \sum_{k=i+1}^{n} \sqrt{M_i M_k}\, \varphi_{k-i}.$$

Summing on i, we have:

$$m_1 + \cdots + m_n \leq \sum_{i=1}^{n} M_i + 4 \sum_{1 \leq i < k \leq n} \sqrt{M_i M_k}\, \varphi_{k-i}$$

$$\leq \sum_{i=1}^{n} M_i + 2 \sum_{1 \leq i < k \leq n} (M_i + M_k)\varphi_{k-i} \leq \theta_n \sum_{i=1}^{n} M_i.$$

Corollary 2.1 then follows from both Theorem 2.4 and the above upper bound.

2.5 New Moment Inequalities

In this section, we derive from Theorem 2.3 new moment inequalities for strongly mixing sequences. These inequalities are similar to the Marcinkiewicz–Zygmund type inequalities for independent random variables. Throughout the section, the strong mixing coefficients are defined in the following way:

$$\alpha_0 = 1/2 \text{ and } \alpha_n = \sup_{k \in \mathbb{Z}} \alpha(\mathcal{F}_k, X_{k+n}) \text{ for any } n > 0. \tag{2.30}$$

Our main result is as follows.

Theorem 2.5 *Let p be any real in $]1, \infty[$. Let $(X_i)_{i \in \mathbb{Z}}$ be a strictly stationary sequence of real-valued random variables with mean 0 and finite moment of order $2p$. Set $Q = Q_{X_0}$. Then, with the notations of Sect. 2.4, for any positive n,*

$$\mathbb{E}(|S_n|^{2p}) \leq (4np)^p \sup_{l \in [1,n]} \mathbb{E}\left(\left|X_0 \sum_{i=0}^{l-1} \mathbb{E}_0(X_i)\right|^p\right). \tag{a}$$

Consequently,

$$\mathbb{E}(|S_n|^{2p}) \leq (8np)^p \int_0^1 [\alpha^{-1}(u) \wedge n]^p Q^{2p}(u)\, du. \tag{b}$$

Remark 2.2 Inequality (a) may be applied to some dynamical systems with hyperbolicity, as shown by Melbourne and Nicol (2008). Inequality (b) can be improved if the strong mixing coefficients are defined by (2.1). We shall obtain Marcinkiewicz–Zygmund type inequalities Marcinkiewicz–Zygmund type inequalities under a weaker mixing condition in Chap. 6 (see Sect. 6.4 and (C.15) in Annex C).

Proof We prove Theorem 2.5 by induction on n. Our induction hypothesis is the following. For any integer $k \leq n$ and any t in $[0, 1]$,

$$\mathbb{E}(|S_{k-1} + tX_k|^{2p}) \leq (4p)^p (k-1+t)^p \sup_{l \in [1,k]} \mathbb{E}\left(\left|X_0 \sum_{i=0}^{l-1} \mathbb{E}_0(X_i)\right|^p\right).$$

First, for any integer $k \leq 4p$,

$$\|S_{k-1} + tX_k\|_{2p} \leq (k-1+t)\|X_0\|_{2p} \leq \sqrt{4p(k-1+t)} \ \|X_0\|_{2p}.$$

Hence the induction hypothesis holds for $k \leq [4p]$.

Now let $n > 4p$. If the induction hypothesis holds at range $n-1$, then, applying Theorem 2.3 with $\psi(x) = |x|^{2p}$, and setting

$$h_n(t) = \mathbb{E}(|S_{n-1} + tX_n|^{2p}) \quad \text{and} \quad \Gamma_n = \sup_{l \in [1,n]} \left\|X_0 \sum_{i=0}^{l-1} \mathbb{E}_0(X_i)\right\|_p,$$

we obtain that

$$\frac{h_n(t)}{4p^2} \leq \sum_{i=1}^{n-1} \int_0^1 \mathbb{E}\left(|S_{i-1} + sX_i|^{2p-2} X_i \sum_{k=i}^n \mathbb{E}_i(X_k)\right) ds$$

$$+ \int_0^t \mathbb{E}(|S_{n-1} + sX_n|^{2p-2} X_n^2) ds.$$

We now apply Hölder's inequality with exponents $p/(p-1)$ and p:

$$\mathbb{E}\left(|S_{i-1} + sX_i|^{2p-2} X_i \sum_{k=i}^n \mathbb{E}_i(X_k)\right) \leq (h_i(s))^{(p-1)/p} \left\|X_i \sum_{k=i}^n \mathbb{E}_i(X_k)\right\|_p.$$

From the stationarity of $(X_i)_{i \in \mathbb{Z}}$,

$$h_n(t) \leq 4p^2 \Gamma_n \left(\sum_{i=1}^{n-1} \int_0^1 (h_i(s))^{(p-1)/p} ds + \int_0^t (h_n(s))^{(p-1)/p} ds\right).$$

Now if the induction hypothesis holds at range $n-1$, then

$$\int_0^1 (h_i(s))^{(p-1)/p} ds \leq (4p\Gamma_n)^{p-1} \int_0^1 (i-1+s)^{p-1} ds$$

$$\leq (4\Gamma_n)^{p-1} p^{p-2} (i^p - (i-1)^p).$$

Set $g_n(s) = (4p(n-1+s)\Gamma_n)^p$. The above inequalities ensure that

$$h_n(t) \leq g_n(0) + 4p^2\Gamma_n \int_0^t (h_n(s))^{(p-1)/p} ds.$$

Now, let

$$H_n(t) = \int_0^t (h_n(s))^{(p-1)/p} ds.$$

The above differential inequality may be written as

$$H_n'(s)(g_n(0) + 4p^2\Gamma_n H_n(s))^{-1+1/p} \leq 1.$$

Integrating this differential inequality between 0 and t yields

$$(h_n(t))^{1/p} - (g_n(0))^{1/p} \leq 4pt\Gamma_n,$$

which implies that $h_n \leq g_n$. Hence (a) holds true.

To prove (b), it is enough to prove that

$$\Gamma_n \leq \|(\alpha^{-1} \wedge n)Q^2\|_p.$$

Let $q = p/(p-1)$. Clearly

$$\Gamma_n \leq \| \sum_{i=0}^{n-1} |\mathbb{E}_0(X_i)| X_0 \|_p.$$

Hence, by the Riesz–Fisher theorem, there exists a random variable Y in $L^q(\mathcal{F}_0)$ such that $\|Y\|_q = 1$ and

$$\Gamma_n \leq \mathbb{E}(Y \sum_{i=0}^{n-1} |X_0 \mathbb{E}_0(X_i)|) \leq \sum_{i=0}^{n-1} \|Y X_0 \mathbb{E}_0(X_i)\|_1.$$

Hence, by (1.11c),

$$\Gamma_n \leq 2 \sum_{i=0}^{n-1} \int_0^{\alpha_i} Q_{YX_0}(u) Q_{X_i}(u) du.$$

Finally, by Lemma 2.1(b)

$$\Gamma_n \leq 2 \int_0^1 Q_Y(u)[\alpha^{-1}(u) \wedge n] Q^2(u) du,$$

which implies (b) via Hölder's inequality on $[0, 1]$ applied to the functions Q_Y and $[\alpha^{-1} \wedge n]Q^2$. ∎

To conclude this section, we give a pseudo exponential inequality for geometrically strongly mixing sequences. Our result is similar to the results of Theorem 6 in Doukhan et al. (1984).

Corollary 2.2 *Let* $(X_i)_{i \in \mathbb{Z}}$ *be a sequence of centered real-valued random variables each bounded a.s. by 1, and* $(\alpha_n)_{n \geq 0}$ *be defined by (2.30). Suppose that, for some* $a < 1$, $\limsup_n \alpha_n^{1/n} < a$. *Then there exists some positive* x_0 *such that, for any* $x \geq x_0$ *and any positive integer* n,

$$\mathbb{P}\left(|S_n| \geq x\sqrt{n \log(1/a)}\right) \leq a^{x/2}.$$

Proof It is easy to check that

$$\limsup_{p \to \infty} p^{-1} \|\alpha^{-1} Q^2\|_p < (-e \log a)^{-1}.$$

Hence there exists some $p_0 > 1$ such that, for any $p \geq p_0$,

$$\|S_n\|_{2p}^2 \leq 4np^2(-e \log a)^{-1}.$$

By the Markov inequality applied to S_n^{2p}, we infer that

$$\mathbb{P}\left(|S_n| \geq x\sqrt{n \log(1/a)}\right) \leq e^{-p}\left(\frac{-2p}{x \log a}\right)^{2p}.$$

Set $p = -(x/2)\log a$. Then the above inequality yields Corollary 2.2 if $p \geq p_0$, which holds true as soon as $x \geq -(2p_0/\log a)$. ∎

Exercises

(1) Let $(X_i)_{i \in \mathbb{Z}}$ be a sequence of centered real-valued random variables with finite fourth moments, and let $(\alpha_n)_{n \geq 0}$ be defined by (2.1).

(a) Let $i \leq j \leq k \leq l$ be natural integers. Prove that

$$|\mathbb{E}(X_i X_j X_k X_l)| \leq 2 \int_0^1 \mathbb{1}_{u < \alpha_{j-i}} \mathbb{1}_{u < \alpha_{l-k}} Q_i(u) Q_j(u) Q_k(u) Q_l(u) du. \qquad (1)$$

(b) Prove that

$$\mathbb{E}(S_n^4) \leq 12 \sum_{1 \leq i \leq j \leq k \leq l \leq n} |\mathbb{E}(X_i X_j X_k X_l)|(1 + \mathbb{1}_{j < k}).$$

(c) Prove that

$$\mathbb{E}(S_n^4) \leq 24 \sum_{j=1}^{n} \sum_{k=1}^{n} \int_0^1 [\alpha^{-1}(u) \wedge n]^2 Q_j^2(u) Q_k^2(u) du. \tag{2}$$

(d) Suppose now that $\|X_k\|_\infty \leq 1$ for any k in $[1, n]$. Derive from the above inequalities that

$$\mathbb{E}(S_n^4) \leq 24n^2 \sum_{m=0}^{n-1} (2m+1)\alpha_m. \tag{3}$$

Compare (3) with (2.13) and (2.11).

(2) Let $(S_n)_{n \geq 0}$ be a martingale sequence in L^p for some $p > 2$ and $X_n = S_n - S_{n-1}$. Either use Inequality (2.3) in Pinelis (1994) or adapt the proof of Theorem 2.5 to prove the inequality (4) below, given in Rio (2009):

$$\|S_n\|_p^2 \leq \|S_0\|_p^2 + (p-1) \sum_{k=1}^{n} \|X_k\|_p^2. \tag{4}$$

Chapter 3
Maximal Inequalities and Strong Laws

3.1 Introduction

In this chapter, we are interested in extensions of the classical maximal inequalities of Kolmogorov and Doob to weakly dependent sequences. Here we adapt previously known tools to the context of weakly dependent sequences. In Sect. 3.2, we give a maximal inequality for second-order moments of the maximum of partial sums. From this maximal inequality we then obtain a criterion for the almost sure convergence of a series of dependent random variables in the style of Kolmogorov's criterion. Next, in Sect. 3.3, we give new maximal inequalities, which are more suitable for long range dependence. These inequalities allow us to get an extension of the results of Berbee (1987) on rates of convergence in the strong law of large numbers for β-mixing sequences to strongly mixing sequences.

3.2 An Extension of the Maximal Inequality of Kolmogorov

Throughout this chapter, $(X_i)_{i\in\mathbb{N}}$ is a sequence of real-valued random variables. The strong mixing coefficients of $(X_i)_{i\in\mathbb{N}}$ are defined by (2.30). We set

$$Q_{X_i} = Q_i, \ S_0 = 0, \ S_k = \sum_{i=1}^{k}(X_i - \mathbb{E}(X_i)) \ \text{and} \ S_n^* = \sup_{k\in[0,n]} S_k. \tag{3.1}$$

In this section, we prove the maximal inequality below.

Theorem 3.1 *Let $(X_i)_{i\in\mathbb{N}}$ be a sequence of centered real-valued random variables with finite variance and λ be any nonnegative real. Set $p_k = \mathbb{P}(S_k^* > \lambda)$. Then*

© Springer-Verlag GmbH Germany 2017
E. Rio, *Asymptotic Theory of Weakly Dependent Random Processes*,
Probability Theory and Stochastic Modelling 80,
DOI 10.1007/978-3-662-54323-8_3

$$\mathbb{E}((S_n^* - \lambda)_+^2) \le 4 \sum_{i=1}^{n} \int_0^{p_i} Q_i(u)\Big(Q_i(u) + 4 \sum_{k=i+1}^{n} Q_k(u)\,\mathbb{1}_{u < \alpha_{k-i}}\Big) du \tag{a}$$

$$\le 16 \sum_{k=1}^{n} \int_0^{p_k} [\alpha^{-1}(u) \wedge n] Q_k^2(u)\,du,$$

with $x_+ = \sup(x, 0)$. *In the particular case $\lambda = 0$,*

$$\mathbb{E}(S_n^{*2}) \le 16 \sum_{k=1}^{n} \int_0^{1} [\alpha^{-1}(u) \wedge n] Q_k^2(u)\,du. \tag{b}$$

From Theorem 3.1(b), one can derive the following extension of Kolmogorov's result on the almost sure convergence of series of random variables. We refer to Cuny and Fan (2016) for more about series of dependent random variables.

Corollary 3.1 *The series $\sum_{i=1}^{\infty} X_i$ converges almost surely as soon as the condition below holds:*

$$\sum_{i=1}^{\infty} \int_0^{1} \alpha^{-1}(u) Q_i^2(u)\,du < +\infty. \tag{a}$$

Application of Corollary 3.1. Suppose that the random variables X_i are defined from a strictly stationary and strongly mixing sequence $(Z_i)_{i \in \mathbb{Z}}$ by $X_i = c_i Z_i$. If Q_{Z_0} satisfies condition (DMR), then condition (a) of Corollary 3.1 holds true as soon as $\sum_{i>0} c_i^2 < \infty$.

Proof of Theorem 3.1 The proof adapts a trick of Garsia (1965) to our context: write

$$(S_n^* - \lambda)_+^2 = \sum_{k=1}^{n} ((S_k^* - \lambda)_+^2 - (S_{k-1}^* - \lambda)_+^2). \tag{3.2}$$

Since $(S_k^*)_{k \ge 0}$ is nondecreasing, the quantities on the right-hand side are nonnegative. Now

$$((S_k^* - \lambda)_+ - (S_{k-1}^* - \lambda)_+)((S_k^* - \lambda)_+ + (S_{k-1}^* - \lambda)_+) > 0$$

if and only if $S_k > \lambda$ and $S_k > S_{k-1}^*$, and then $S_k = S_k^*$. Consequently

$$(S_k^* - \lambda)_+^2 - (S_{k-1}^* - \lambda)_+^2 \le 2(S_k - \lambda)((S_k^* - \lambda)_+ - (S_{k-1}^* - \lambda)_+), \tag{3.3}$$

which implies that

$$(S_n^* - \lambda)_+^2 \le 2 \sum_{k=1}^{n} (S_k - \lambda)(S_k^* - \lambda)_+ - 2 \sum_{k=1}^{n} (S_k - \lambda)(S_{k-1}^* - \lambda)_+$$

$$\leq 2(S_n - \lambda)_+(S_n^* - \lambda)_+ - 2\sum_{k=1}^{n}(S_{k-1}^* - \lambda)_+ X_k. \tag{3.4}$$

Since

$$(S_n - \lambda)_+(S_n^* - \lambda)_+ \leq \frac{1}{4}(S_n^* - \lambda)_+^2 + (S_n - \lambda)_+^2,$$

it follows that

$$(S_n^* - \lambda)_+^2 \leq 4(S_n - \lambda)_+^2 - 4\sum_{k=1}^{n}(S_{k-1}^* - \lambda)_+ X_k. \tag{3.5}$$

Next we bound $(S_n - \lambda)_+^2$. Adapting the decomposition (3.2), we get:

$$(S_n - \lambda)_+^2 = \sum_{k=1}^{n}((S_k - \lambda)_+^2 - (S_{k-1} - \lambda)_+^2)$$

$$= 2\sum_{k=1}^{n}(S_{k-1} - \lambda)_+ X_k + 2\sum_{k=1}^{n}X_k^2 \int_0^1 (1-t)\,\mathbb{1}_{S_{k-1}+tX_k>\lambda}dt. \tag{3.6}$$

Noticing then that $\mathbb{1}_{S_{k-1}+tX_k>\lambda} \leq \mathbb{1}_{S_k^*>\lambda}$, we infer from (3.6) that

$$(S_n - \lambda)_+^2 \leq 2\sum_{k=1}^{n}(S_{k-1} - \lambda)_+ X_k + \sum_{k=1}^{n}X_k^2\,\mathbb{1}_{S_k^*>\lambda}.$$

From (3.5) and the above inequality we now obtain that

$$(S_n^* - \lambda)_+^2 \leq 4\sum_{k=1}^{n}(2(S_{k-1} - \lambda)_+ - (S_{k-1}^* - \lambda)_+)X_k + 4\sum_{k=1}^{n}X_k^2\,\mathbb{1}_{S_k^*>\lambda}. \tag{3.7}$$

Set $D_0 = 0$ and $D_k = 2(S_k - \lambda)_+ - (S_k^* - \lambda)_+$ for any positive k. Clearly

$$\text{Cov}(D_{k-1}, X_k) = \sum_{i=1}^{k-1}\text{Cov}(D_i - D_{i-1}, X_k).$$

Now the random variables $D_i - D_{i-1}$ are measurable with respect to $\mathcal{F}_i = \sigma(X_j : j \leq i)$. Hence

$$\mathbb{E}((S_n^* - \lambda)_+^2) \leq 4\sum_{k=1}^{n}\mathbb{E}(X_k^2\,\mathbb{1}_{S_k^*>\lambda}) + 4\sum_{i=1}^{n-1}\mathbb{E}\left(\left|(D_i - D_{i-1})\sum_{k=i+1}^{n}\mathbb{E}_i(X_k)\right|\right).$$

$$\tag{3.8}$$

In order to bound $Q_{D_i-D_{i-1}}$, we now bound $|D_i - D_{i-1}|$. If $(S_i^* - \lambda)_+ = (S_{i-1}^* - \lambda)_+$, then

$$|D_i - D_{i-1}| = 2|(S_i - \lambda)_+ - (S_{i-1} - \lambda)_+| \leq 2|X_i| \, \mathbb{I}_{S_i^* > \lambda},$$

since $D_i - D_{i-1} = 0$ if $S_i \leq \lambda$ and $S_{i-1} \leq \lambda$. In the opposite case $S_i = S_i^* > \lambda$ and $S_{i-1} \leq S_{i-1}^* < S_i$. Thus $D_i - D_{i-1} = (S_i - \lambda)_+ + (S_{i-1}^* - \lambda)_+ - 2(S_{i-1} - \lambda)_+$ belongs to $[0, 2|(S_i - \lambda)_+ - (S_{i-1} - \lambda)_+|]$. In each case $|D_i - D_{i-1}| \leq 2|X_i| \, \mathbb{I}_{S_i^* > \lambda}$, whence

$$\mathbb{E}((S_n^* - \lambda)_+^2) \leq 4 \sum_{k=1}^{n} \mathbb{E}(X_k^2 \mathbb{I}_{S_k^* > \lambda}) + 8 \sum_{i=1}^{n-1} \mathbb{E}\left(\mathbb{I}_{S_i^* > \lambda} |X_i \sum_{k=i+1}^{n} \mathbb{E}_i(X_k)| \right). \quad (3.9)$$

Next, by (1.11c) together with Lemma 2.1,

$$\mathbb{E}\left(\mathbb{I}_{S_i^* > \lambda} |X_i \mathbb{E}_i(X_k)| \right) \leq 2 \int_0^{\alpha_{k-i}} Q_i(u) Q_k(u) \mathbb{I}_{u < p_i} du$$

$$\leq \int_0^{\alpha_{k-i}} (Q_i^2(u) \mathbb{I}_{u < p_i} + Q_k^2(u) \mathbb{I}_{u < p_k}) du \quad (3.10)$$

(use the monotonicity of the sequence $(p_k)_{k>0}$). Now, by Lemma 2.1(a), we also have

$$\mathbb{E}(X_k^2 \mathbb{I}_{S_k^* > \lambda}) \leq \int_0^{p_k} Q_k^2(u) du.$$

Theorem 3.1 then follows from (3.9), (3.10) and the above inequality. ■

3.3 Rates of Convergence in the Strong Law of Large Numbers

Let r be any real in $]1, 2[$ and $(X_i)_{i \in \mathbb{N}}$ be a strictly stationary sequence. Theorem 3.1 applied to the sequence $(X_i)_{i \in \mathbb{N}}$ provides the almost sure convergence of $n^{-1/r} S_n$ to 0 under the strong mixing condition

$$\int_0^1 \alpha^{-1}(u) Q_{X_0}^r(u) du < \infty.$$

This condition needs the summability of the series of strong mixing coefficients, even for bounded random variables. By contrast, for strictly stationary and β-mixing sequences of bounded random variables, Berbee (1987) proved the almost sure convergence of $n^{-1/r} S_n$ to 0 under the β-mixing condition

$$\sum_{i \geq 0} (i + 1)^{r-2} \beta_i < \infty, \qquad (BER)$$

which is clearly weaker than the summability of β-mixing coefficients. In the strong mixing case, Shao (1993) has given some rates of convergence in the strong law of large numbers under weaker conditions than the integral condition above. However, in the bounded case, he does not obtain the convergence of $n^{-1/r} S_n$ to 0 under the strong mixing condition corresponding to Berbee's condition. In this section, we give a new maximal inequality, which minimizes the effects of long range dependence. This inequality is then applied to obtain rates of convergence in the strong law of large numbers under minimal assumptions on the mixing coefficients and the tails of the random variables, as in Rio (1995a).

Theorem 3.2 *Let $(X_i)_{i \in \mathbb{N}}$ be a sequence of centered random variables with finite variances. Then, for any nonnegative integer p and any positive x,*

$$\mathbb{P}(S_n^* \geq 2x) \leq \frac{4}{x^2} \sum_{k=1}^{n} \int_0^1 [\alpha^{-1}(u) \wedge p] Q_k^2(u) du + \frac{4}{x} \sum_{k=1}^{n} \int_0^{\alpha_p} Q_k(u) du.$$

Before proving Theorem 3.2, we give an application to the rates of convergence in the strong law. We refer to Dedecker and Merlevède (2007) for extensions of the corollary below to other types of dependence and Banach-valued random variables.

Corollary 3.2 *Let $(X_i)_{i \in \mathbb{N}}$ be a sequence of centered and integrable real-valued random variables. Set $Q = \sup_{i>0} Q_i$.*

(i) Let r be any real in $]1, 2[$. Suppose that

$$M_{r,\alpha}(Q) = \int_0^1 [\alpha^{-1}(u)]^{r-1} Q^r(u) du < +\infty. \qquad (a)$$

Then $n^{-1/r} S_n$ converges to 0 almost surely.
(ii) Suppose that Q satisfies the weaker condition

$$\int_0^1 Q(u) \log(1 + \alpha^{-1}(u)) du < \infty. \qquad (b)$$

Then $n^{-1} S_n$ converges almost surely to 0.

Remark 3.1 Let X be a nonnegative random variable such that $Q_X = Q$. For m-dependent sequences conditions (a) and (b) are respectively equivalent to the usual integrability conditions $\mathbb{E}(X^r) < \infty$ and $\mathbb{E}(X) < \infty$. Note that, in the stationary and ergodic case, the strong law of large numbers holds as soon as the variables are integrable. This result does not hold for non-stationary strongly mixing sequences: condition (b) cannot be relaxed, as proved in Rio (1995a).

Remark 3.2 We refer to Annex C for more about conditions (a) and (b). Notice that (a) and (b) are respectively equivalent to the condition below with r in $]1, 2[$ or $r = 1$:

$$\sum_{i \geq 0}(i + 1)^{r-2} \int_0^{\alpha_i} Q^r(u)du < \infty. \tag{3.11}$$

For bounded sequences, (3.11) is equivalent to the strong mixing condition

$$\sum_{i \geq 0}(i + 1)^{r-2}\alpha_i < \infty.$$

Since $\alpha_i \leq \beta_i$, Corollary 3.2 includes Berbee's result.

Proof of Theorem 3.2 Dividing the random variables by x if necessary, we may assume that $x = 1$. Define the function g by $g(y) = y - 1$ for y in $[1, 2]$, $g(y) = 0$ for $y \leq 1$ and $g(y) = 1$ for $y \geq 2$. Then

$$\mathbb{P}(S_n^* \geq 2) \leq \mathbb{E}(g(S_n^*)) \leq \sum_{k=1}^n \mathbb{E}(g(S_k^*) - g(S_{k-1}^*)).$$

Let f be the nonnegative and differentiable function defined by $f(y) = y^2$ for y in $[0, 1]$, $f(y) = 2y - 1$ for $y \geq 1$ and $f(y) = 0$ for any negative y. Since g is nondecreasing, $g(S_k^*) - g(S_{k-1}^*) \geq 0$. If this quantity is strictly positive, then $S_k > S_{k-1}^*$ and $S_k > 1$. Hence

$$g(S_k^*) - g(S_{k-1}^*) \leq (g(S_k^*) - g(S_{k-1}^*))(2S_k - 1),$$

which implies that

$$\mathbb{P}(S_n^* \geq 2) \leq \sum_{k=1}^n \mathbb{E}\Big((g(S_k^*) - g(S_{k-1}^*))(2S_k - 1)\Big)$$

$$\leq \mathbb{E}((2S_n - 1)g(S_n^*)) - 2\sum_{k=1}^n \mathrm{Cov}(g(S_{k-1}^*), X_k)$$

$$\leq \mathbb{E}(f(S_n)g(S_n^*)) - 2\sum_{k=1}^n \mathrm{Cov}(g(S_{k-1}^*), X_k)$$

$$\leq \mathbb{E}(f(S_n)) - 2\sum_{k=1}^n \mathrm{Cov}(g(S_{k-1}^*), X_k). \tag{3.12}$$

Now, since f' is 2-Lipschitz,

$$\mathbb{E}(f(S_n)) = \sum_{k=1}^{n} \mathbb{E}(f(S_k) - f(S_{k-1})) \le \sum_{k=1}^{n} \operatorname{Var} X_k + \sum_{k=1}^{n} \operatorname{Cov}(f'(S_{k-1}), X_k).$$

(3.13)

Set $g_k(X_1, \ldots X_{k-1}) = \frac{1}{2}f'(S_{k-1}) - g(S^*_{k-1})$. From (3.12) and (3.13) we get that

$$\mathbb{P}(S^*_n \ge 2) \le \sum_{k=1}^{n} \operatorname{Var} X_k + 2 \sum_{k=1}^{n} \operatorname{Cov}(g_k(X_1, \ldots X_{k-1}), X_k).$$ (3.14)

Recall that $\frac{1}{2}f'$ and g are 1-Lipschitz and coordinatewise nondecreasing. Therefore the function g_k is separately 1-Lipschitz with respect to each coordinate. For $i \le k-1$, let

$$D'_{i,k} = g_k(X_1, \ldots, X_i, 0, \ldots, 0) - g_k(X_1, \ldots, X_{i-1}, 0, \ldots, 0).$$

For any nonnegative integer p,

$$g_k(X_1, \ldots X_{k-1}) = g_k(X_1, \ldots, X_{(k-p)^+}, 0, \ldots, 0) + \sum_{i=(k-p)_++1}^{k-1} D'_{i,k}. \quad (3.15)$$

Now the first term on the right vanishes if $p \ge k$. Since g_k has values in $[-1, 1]$ and the first term on the right is measurable with respect to $\sigma(X_i : i \le k - p)$, by Theorem 1.1(a),

$$|\operatorname{Cov}(g_k(X_1, \ldots, X_{(k-p)^+}, 0, \ldots, 0), X_k) \le 2 \int_0^{\alpha_p} Q_k(u)du.$$

Next the random variables $D'_{i,k}$ satisfy $|D'_{i,k}| \le |X_{i-k}|$ and are measurable with respect to $\sigma(X_i : i \le k - i)$, whence

$$|\operatorname{Cov}(D'_{i,k}, X_k)| \le 2 \int_0^{\alpha_i} Q_{k-i}(u)Q_k(u)du \le \int_0^{\alpha_i} (Q^2_{k-i}(u) + Q^2_k(u))du.$$

Both (3.15) and the above two inequalities ensure that

$$|\operatorname{Cov}(g_k(X_1, \ldots X_{k-1}), X_k)| \le \sum_{i=1}^{p-1} \int_0^{\alpha_i} (Q^2_{k-i}(u) + Q^2_k(u))du + 2 \int_0^{\alpha_p} Q_k(u)du.$$

(3.16)

Now, both (3.14), (3.16) and the elementary inequality $\operatorname{Var} X_k \le 2 \int_0^{\alpha_0} Q^2_k(u)du$ imply Theorem 3.2. ∎

Proof of Corollary 3.2 The proof of Corollary 3.2 is a direct consequence of Proposition 3.1 below applied to the sequences $(X_i)_{i\in\mathbb{N}}$ and $(-X_i)_{i\in\mathbb{N}}$ via the Borel–Cantelli lemma: indeed the series in Proposition 3.1 are convergent if and only if for any positive ε,

$$\sum_{N>0} \mathbb{P}(S_{2^N}^* > \varepsilon 2^{N/r}) < \infty,$$

which implies the convergence of $n^{-1/r} S_n^*$ to 0, due to the monotonicity of S_n^*.

Proposition 3.1 *With the same notation as in Theorem 3.2, under condition (a) of Corollary 3.2, for any positive ε,*

$$\sum_{n>0} n^{-1}\mathbb{P}(S_n^* \geq \varepsilon n^{1/r}) < \infty. \tag{a}$$

Under condition (b) of Corollary 3.2, for any positive ε,

$$\sum_{n>0} n^{-1}\mathbb{P}(S_n^* \geq \varepsilon n) < \infty. \tag{b}$$

Proof For arbitrary v in $[0, 1]$, let the sequences $(\bar{X}_i)_{i\in\mathbb{Z}}$ and $(\tilde{X}_i)_{i\in\mathbb{Z}}$ be defined by

$$\bar{X}_i = (X_i \wedge Q(v)) \vee (-Q(v)) \text{ and } \tilde{X}_i = X_i - \bar{X}_i.$$

Let U be uniformly distributed over $[0, 1]$. Since $|X_i|$ has the same distribution as $Q_i(U)$,

$$Q_{\bar{X}_i}(u) = Q_i(u) \wedge Q(v) \text{ and } Q_{\tilde{X}_i}(u) = \sup(Q_i(u) - Q(v), 0).$$

Now $Q_i \leq Q$, whence

$$|\mathbb{E}(\tilde{X}_i)| \leq \mathbb{E}(|\tilde{X}_i|) \leq \int_0^v (Q(u) - Q(v))du. \tag{3.17}$$

Set $\bar{S}_k = \sum_{i=1}^k (\bar{X}_i - \mathbb{E}(\bar{X}_i))$ and $\bar{S}_n^* = \sup_{k\in[0,n]} \bar{S}_k$. Noting that

$$S_n^* \leq \bar{S}_n^* + \sum_{i=1}^n (|\tilde{X}_i| + |\mathbb{E}(\tilde{X}_i)|),$$

we infer from (3.17) that

$$n^{-1}\mathbb{P}(S_n^* \geq 5x) \leq n^{-1}\mathbb{P}(\bar{S}_n^* \geq 4x) + \frac{2}{x}\int_0^v (Q(u) - Q(v))du. \tag{3.18}$$

Next, by Theorem 3.2 applied to the random variables $\bar{X}_i$, we get that

$$n^{-1}\mathbb{P}(\bar{S}_n^* \geq 4x) \leq \frac{1}{x^2} \int_0^1 [\alpha^{-1}(u) \wedge p]Q^2(v \vee u)du + \frac{2}{x} \int_0^{\alpha_p} Q(v \vee u)du.$$

(3.19)

We now choose the parameters p and v in such a way that the terms on the right-hand side lead to the same integral condition. Set

$$R(u) = \alpha^{-1}(u)Q(u).$$

(3.20)

In the strong mixing case, R plays the same role as the quantile function Q in the independent case. We will choose v in such a way that $R(v)$ is of the order of $n^{1/r}$. Before choosing v, we choose $p = \alpha^{-1}(v)$, in order to get upper bounds of the same order of magnitude in (3.18) and (3.19). With this choice of p, $\alpha_p \leq v$. Consequently,

$$\int_0^1 [\alpha^{-1}(u) \wedge p]Q^2(v \vee u)du \leq \int_0^1 R(v \vee u)Q(u)du.$$

Therefore, by (3.18) and (3.19),

$$n^{-1}\mathbb{P}(S_n^* \geq 5x) \leq \frac{2}{x} \int_0^v Q(u)du + \frac{1}{x^2} \int_0^1 R(v \vee u)Q(u)du.$$

(3.21)

Let ε be any real in $]0, 1]$. Set $x = x_n = \varepsilon n^{1/r}$ and $v = v_n = R^{-1}(n^{1/r})$ in (3.21). Since R is right-continuous and nonincreasing,

$$(R(u) \leq n^{1/r}) \text{ if and only if } (u \geq v_n),$$

(3.22)

whence

$$\int_0^{v_n} R(v_n)Q(u)du \leq n^{1/r} \int_0^{v_n} Q(u)du.$$

It follows that

$$n^{-1}\mathbb{P}(S_n^* \geq 5x_n) \leq 3\varepsilon^{-2}\left(n^{-1/r} \int_0^{v_n} Q(u)du + n^{-2/r} \int_{v_n}^1 R(u)Q(u)du\right).$$

(3.23)

Let us prove (a). Set $c_\varepsilon = \varepsilon^2/3$. Summing on n the inequalities (3.23), we get that

$$c_\varepsilon \sum_{n>0} \frac{1}{n}\mathbb{P}(S_n^* \geq 5x_n) \leq \int_0^1 Q(u) \sum_{n>0}\left(n^{-1/r}\,\mathbb{1}_{u<v_n} + n^{-2/r}R(u)\,\mathbb{1}_{u\geq v_n}\right)du,$$

with $x_n = \varepsilon n^{1/r}$. Now, applying (3.22),

$$\sum_{n>0}\frac{c_\varepsilon}{n}\mathbb{P}(S_n^* \geq 5x_n) \leq \int_0^1 Q(u)\sum_{n>0}\left(n^{-1/r}\,1\!\!1_{n<R^r(u)} + n^{-2/r}R(u)\,1\!\!1_{n\geq R^r(u)}\right)du.$$

$$(3.24)$$

Since r belongs to $]1,2[$, there exist constants c_r and C_r depending only on r such that

$$\sum_{0<n<R^r(u)} n^{-1/r} \leq c_r R^{r-1}(u) \quad\text{and}\quad \sum_{n\geq R^r(u)\vee 1} n^{-2/r} \leq C_r(R^{r-2}(u) \wedge 1).$$

Both the above inequalities and (3.24) ensure that

$$\sum_{n>0}\frac{1}{n}\mathbb{P}(S_n^* \geq 5\varepsilon n^{1/r}) \leq C\int_0^1 R^{r-1}(u)Q^r(u)du,$$

for some constant C depending only on r, which completes the proof of Proposition 3.1(a).

To prove Proposition 3.1(b), we need to truncate the random variables again. Let

$$Y_i = (X_i \wedge n) \vee (-n), \quad \tilde{Y}_i = X_i - Y_i \quad\text{and}\quad T_n^* = \sup_{k\in[0,n]}\sum_{i=1}^k (Y_i - \mathbb{E}(Y_i)).$$

Since $Q_{Y_i} \leq Q \wedge n$ for any i in $[1,n]$, it follows from (3.23) that

$$n^{-1}\mathbb{P}(T_n^* \geq 5\varepsilon n) \leq c_\varepsilon^{-1}\left(n^{-1}\int_0^{v_n}(Q(u)\wedge n)du + n^{-2}\int_{v_n}^1 R(u)Q(u)du\right).$$

$$(3.25)$$

Now set $\Gamma = \bigcup_{i=1}^n (X_i \neq Y_i)$. For any $\omega \notin \Gamma$,

$$S_n^*(\omega) \leq T_n^*(\omega) + \sum_{i=1}^n \mathbb{E}(|Y_i - X_i|). \tag{3.26}$$

Let X be a nonnegative random variable such that $Q_X = Q$. Then

$$\mathbb{P}(\Gamma) \leq \sum_{i=1}^n \mathbb{P}(|X_i| > n) \leq n\mathbb{P}(X > n)$$

and

$$\sum_{i=1}^n \mathbb{E}(|Y_i - X_i|) = \sum_{i=1}^n \int_n^\infty \mathbb{P}(|X_i| > u)du \leq n\mathbb{E}((X-n)_+).$$

Since $\mathbb{E}((X - n)_+) \leq \varepsilon$ for n large enough, there exists some positive integer n_0 such that, for any $n \geq n_0$,

$$\frac{c_\varepsilon}{n}\mathbb{P}(S_n^* \geq 6\varepsilon n) \leq \mathbb{P}(X > n) + \left(\frac{1}{n}\int_0^{v_n}(Q(u) \wedge n)du + \frac{1}{n^2}\int_{v_n}^1 R(u)Q(u)du\right).$$

(3.27)

Set $w_n = Q^{-1}(n) = \mathbb{P}(X > n)$. Since $w_n \leq u < v_n$ if and only if $Q(u) \leq n < R(u)$, we get that

$$n^{-1}\int_0^{v_n}(Q(u) \wedge n)du = \mathbb{P}(X > n) + n^{-1}\int_0^1 Q(u)\,\mathrm{1\!\!I}_{Q(u)\leq n<R(u)}du.$$

Hence for $n \geq n_0$,

$$\frac{c_\varepsilon}{n}\mathbb{P}(S_n^* \geq 6\varepsilon n) \leq \mathbb{P}(X > n) + \int_0^1 Q(u)\left(\frac{1}{n}\,\mathrm{1\!\!I}_{Q(u)\leq n<R(u)} + \frac{R(u)}{n^2}\,\mathrm{1\!\!I}_{n\geq R(u)}\right)du.$$

(3.28)

Finally, since

$$\sum_{Q(u)\vee 1\leq n<R(u)} n^{-1} \leq 1 + \log(1 + \alpha^{-1}(u)) \text{ and } \sum_{n\geq R(u)\vee 1} n^{-2} \leq 2(R(u) \vee 1)^{-1},$$

(3.28) implies Proposition 3.1(b). ■

Exercises

(1) Let $(X_i)_{i\in\mathbb{N}}$ be a sequence of real-valued and integrable centered random variables. Set $Q = \sup_{i>0} Q_i$ and let R be defined by (3.20).
  (a) Prove that, for any positive $x > 0$,

$$n^{-1}\mathbb{P}(S_n^* \geq 5x) \leq \frac{3}{x}\int_0^{R^{-1}(x)} Q(u)du + \frac{1}{x^2}\int_{R^{-1}(x)}^1 R(u)Q(u)du.$$

(1)

Hint: apply (3.21).
  (b) Prove that, for any r in $]1, 2[$,

$$\mathbb{E}(S_n^{*r}) = r5^r\int_0^\infty x^{r-1}\mathbb{P}(S_n^* \geq 5x)dx.$$

(2)

  (c) Infer from (2) that

$$\mathbb{E}(S_n^{*r}) \le n5^r \frac{r(5-2r)}{(r-1)(2-r)} \int_0^1 [\alpha^{-1}(u)]^{r-1} Q^r(u) du. \tag{3}$$

Prove that (3) still holds if $\alpha^{-1}(u)$ is changed to $\alpha^{-1}(u) \wedge n$.

(2) Let $(X_i)_{i \in \mathbb{N}}$ be a sequence of real-valued and integrable centered random variables. Assume that $\mathbb{E}(|X_i|^p) < \infty$ for any positive i, for some fixed $p > 2$.

(a) Let S be a nonnegative random variable. Prove that

$$2\mathbb{E}(S^p) = p(p-1)(p-2) \int_0^\infty \mathbb{E}((S-\lambda)_+^2) \lambda^{p-3} d\lambda. \tag{4}$$

(b) Set $H(\lambda) = \mathbb{P}(S_n^* > \lambda)$. Prove that

$$\mathbb{E}(S_n^{*p}) \le 8p(p-1)(p-2) \sum_{k=1}^n \int_0^\infty \left(\int_0^{H(\lambda)} [\alpha^{-1}(u) \wedge n] Q_k^2(u) du \right) \lambda^{p-3} d\lambda. \tag{5}$$

(c) Starting from (5), prove that

$$\mathbb{E}(S_n^{*p}) \le 8p(p-1) \sum_{k=1}^n \int_0^1 [\alpha^{-1}(u) \wedge n] Q_k^2(u) Q_{S_n^*}^{p-2}(u) du. \tag{6}$$

Hint: apply Fubini's Theorem and note that the inverse function of H is $Q_{S_n^*}$.

(d) Prove that

$$\mathbb{E}(S_n^{*p}) \le [8p(p-1)]^{p/2} n^{(p-2)/2} \sum_{k=1}^n \int_0^1 [\alpha^{-1}(u) \wedge n]^{p/2} Q_k^p(u) du. \tag{7}$$

Compare (7) with the inequalities of Chap. 2 and with Corollary 1 in Yokoyama (1980).

(3) *A Marcinkiewicz–Zygmund inequality for martingales*. Let $(X_i)_{i \in \mathbb{N}}$ be a sequence of real-valued and integrable centered random variables and let $\mathcal{F}_k = \sigma(X_i : i \le k)$. Suppose that $(S_k)_{k \ge 0}$ is a martingale with respect to $\mathcal{F}_k$. Let $p > 2$. Prove that, if $\mathbb{E}(|X_i|^p) < \infty$ for any positive i, then

$$\mathbb{E}(S_n^{*p}) \le [4p(p-1)]^{p/2} n^{(p-2)/2} \sum_{i=1}^n \mathbb{E}(|X_i|^p). \tag{8}$$

Hint: apply (3.8) and use the ideas of Exercise 2.

(4) *A maximal inequality of Serfling*. In this exercise, we prove an inequality of Serfling (1970) in a particular case. Let $p > 2$ and $(X_i)_{i \in \mathbb{N}}$ be a sequence of real-valued random variables such that for any pair of natural numbers (m, n) such that $m < n$,

$$\mathbb{E}((S_n - S_m)_+^p) \le (n-m)^{p/2}. \tag{9}$$

The goal of the exercise is to prove that there exists some constant $K(p)$ such that

$$\mathbb{E}(S_n^{*p}) \leq K(p)n^{p/2}. \tag{10}$$

(a) Let

$$\varphi(N) = \sup_{k \geq 0} \| \sup_{i \in [0, 2^N]} S_{k2^N + i} - S_{k2^N} \|_p.$$

Prove that, for any positive integer N,

$$\varphi(N) \leq \sup_{k \geq 0} \|(S_{(k+1/2)2^N} - S_{k2^N})_+\|_p + 2^{1/p} \varphi(N-1). \tag{11}$$

(b) Infer that

$$\varphi(N) \leq (2^{1/2} - 2^{1/p})^{-1} 2^{N/2}.$$

Next prove that (10) holds true with $K(p) = (1 - 2^{\frac{1}{p} - \frac{1}{2}})^{-p}$. Compare (10) with the Doob–Kolmogorov inequality for martingales.

Chapter 4
Central Limit Theorems

4.1 Introduction

In this chapter, we are interested in the convergence in distribution of suitably normalized partial sums of a strongly mixing and stationary sequence of real-valued random variables. In Sect. 4.2, we give the extension of Ibragimov's central limit theorem for partial sums of a sstrongly mixing sequence of bounded random variables to unbounded random variables, due to Doukhan et al. (1994). We essentially follow the line of proof of Ibragimov and Linnik (1971) and Hall and Heyde (1980). This approach is based on Gordin's theorem (1969) on martingale approximation. Next, in Sect. 4.3, we prove a functional central limit theorem for the normalized partial sum process under the same integrability condition on the tails of the random variables. In Sect. 4.4, we give a triangular version of the central limit theorem. This result is obtained by adapting the Lindeberg method to the dependent case.

4.2 A Central Limit Theorem for Strongly Mixing and Stationary Sequences

In this section, we derive a central limit theorem for partial sums from the covariance inequalities of Chap. 1. Our proof is based on Theorem 5.2 in Hall and Heyde (1980), which is a consequence of Gordin's results (1969, 1973) on approximation by martingales (see Volný (1993) for a survey). We first recall Theorem 5.2 in Hall and Heyde.

Theorem 4.1 *Let $(X_i)_{i\in\mathbb{Z}}$ be a stationary and ergodic sequence of real-valued random variables and let*

$$S_n = \sum_{i=1}^{n}(X_i - \mathbb{E}(X_i)) \ \text{and} \ \mathcal{F}_0 = \sigma(X_i : i \leq 0).$$

© Springer-Verlag GmbH Germany 2017
E. Rio, *Asymptotic Theory of Weakly Dependent Random Processes*,
Probability Theory and Stochastic Modelling 80,
DOI 10.1007/978-3-662-54323-8_4

Suppose that, for any nonnegative integer n,

$$\sum_{k>0} \text{Cov}(\mathbb{E}(X_n \mid \mathcal{F}_0), X_k) \ \text{converges} \tag{a}$$

and

$$\lim_{n\to+\infty} \sup_{K>0} \left| \sum_{k\geq K} \text{Cov}(\mathbb{E}(X_n \mid \mathcal{F}_0), X_k) \right| = 0. \tag{b}$$

Then $n^{-1} \text{Var} S_n$ converges to a nonnegative real σ^2 and $n^{-1/2} S_n$ converges in distribution to the normal law $N(0, \sigma^2)$.

The proof of Theorem 4.1 can be found in Hall and Heyde (1980). Now we derive from Theorem 4.1 a central limit theorem for partial sums of a strongly mixing sequence.

Theorem 4.2 *Let $(X_i)_{i\in\mathbb{Z}}$ be a strictly stationary and ergodic sequence of real-valued random variables satisfying condition (DMR) of Corollary 1.2, with the strong mixing coefficients defined by (2.30). Then $n^{-1} \text{Var} S_n$ converges to a nonnegative real σ^2 and $n^{-1/2} S_n$ converges in distribution to the normal law $N(0, \sigma^2)$.*

Remark 4.1 From Lemma 1.1, σ^2 is equal to the series of covariances defined in Lemma 1.1. It is worth noting that Theorem 4.2 implies the uniform integrability of the sequence of random variables $(n^{-1} S_n^2)_{n>0}$. This fact follows from Theorem 5.4 in Billingsley (1968). We refer to Merlevède and Peligrad (2000) for a central limit theorem under a weaker strong mixing condition.

If the strong mixing coefficients $(\alpha_n)_{n\geq 0}$ are defined by (2.1), then the convergence of α_n to 0 implies the ergodicity of the sequence, and consequently the ergodicity assumption can be removed. If the strong mixing coefficients are defined by (2.30), the ergodicity assumption cannot be removed, as proved by the counterexample of Exercise 1, Chap. 4. In Sect. 9.7 of Chap. 9, we will prove the optimality of condition (DMR) for power type mixing rates. We mention that Bradley (1997) shows that condition (DMR) is optimal for arbitrary mixing rates.

Proof of Theorem 4.2. We have to prove that assumptions (a) and (b) of Theorem 4.1 hold true under condition (DMR), for an ergodic sequence. Clearly these conditions are implied by the absolute convergence of the series of covariance together with the condition

$$\lim_{n\to+\infty} \sum_{k>0} |\text{Cov}(\mathbb{E}(X_n \mid \mathcal{F}_0), X_k)| = 0. \tag{4.1}$$

To prove that (4.1) holds true, we now apply Theorem 1.1(a). Let

$$X_n^0 = \mathbb{E}(X_n \mid \mathcal{F}_0). \tag{4.2}$$

Since the random variable X_n^0 is $\mathcal{F}_0$-measurable, Theorem 1.1(a) yields

$$|\mathrm{Cov}(\mathbb{E}(X_n \mid \mathcal{F}_0), X_k)| \le 2 \int_0^{\alpha_k} Q_{X_n^0}(u) Q_{X_k}(u) du. \tag{4.3}$$

Let δ be a random variable with uniform distribution over $[0, 1]$, independent of $(X_i)_{i \in \mathbb{Z}}$. Then (see Exercise 1, Chap. 1) the random variable

$$U_n^0 = H_{X_n^0}(|X_n^0|) + \delta(H_{X_n^0}(|X_n^0| - 0) - H_{X_n^0}(|X_n^0|))$$

has the uniform distribution over $[0, 1]$ and $Q_{X_n^0}(U_n^0) = |X_n^0|$ almost surely. Let ε_n^0 denote the sign of X_n^0. Then

$$\int_0^{\alpha_k} Q_{X_n^0}(u) Q_{X_k}(u) du = \mathbb{E}(X_n^0 \varepsilon_n^0 Q_{X_k}(U_n^0) \mathbb{I}_{U_n^0 \le \alpha_k}). \tag{4.4}$$

Since δ is independent of $(X_i)_{i \in \mathbb{Z}}$,

$$X_n^0 = \mathbb{E}(X_n \mid \sigma(\delta) \vee \mathcal{F}_0),$$

whence

$$\int_0^{\alpha_k} Q_{X_n^0}(u) Q_{X_k}(u) du = \mathbb{E}(X_n \varepsilon_n^0 Q_{X_k}(U_n^0) \mathbb{I}_{U_n^0 \le \alpha_k}).$$

It follows that

$$\int_0^{\alpha_k} Q_{X_n^0}(u) Q_{X_k}(u) du \le \mathbb{E}(|X_n| Q_{X_k}(U_n^0) \mathbb{I}_{U_n^0 \le \alpha_k}). \tag{4.5}$$

Now, by (4.5) and Lemma 2.1(a),

$$\int_0^{\alpha_k} Q_{X_n^0}(u) Q_{X_k}(u) du \le \int_0^{\alpha_k} Q_{X_n}(u) Q_{X_k}(u) du. \tag{4.6}$$

Both (4.3) and (4.6) together with the stationarity of the sequence then ensure that

$$|\mathrm{Cov}(X_n^0, X_k)| \le 2 \int_0^{\alpha_k} Q_{X_0}^2(u) du. \tag{4.7}$$

Now

$$\mathrm{Cov}(X_n^0, X_k) = \mathrm{Cov}(X_n^0, X_k^0) = \mathrm{Cov}(X_k^0, X_n) \tag{4.8}$$

and therefore, interchanging k and n in (4.7), we get that

$$|\mathrm{Cov}(X_n^0, X_k)| \le 2 \int_0^{\alpha_n} Q_{X_0}^2(u) du. \tag{4.9}$$

Consequently,

$$|\text{Cov}(X_n^0, X_k)| \le 2 \int_0^{\inf(\alpha_k, \alpha_n)} Q_{X_0}^2(u) du,$$

which ensures the normal convergence of the series. Hence (4.1) holds true, which completes the proof of Theorem 4.2. ∎

Starting from Theorem 4.2, we now derive a central limit theorem for stationary and strongly mixing sequences of random variables in the multivariate case.

Corollary 4.1 *Let $(X_i)_{i \in \mathbb{Z}}$ be a strictly stationary sequence of random variables with values in $\mathbb{R}^d$. Let Q_0 be the generalized inverse function of $H_{X_0}(t) = \mathbb{P}(\|X_0\| > t)$. Assume that $(X_i)_{i \in \mathbb{Z}}$ satisfies condition (DMR) of Corollary 1.2, with the strong mixing coefficients defined by (2.1). Then $n^{-1}\text{Cov}(S_n, S_n)$ converges to $\Gamma = \text{Var} X_0 + 2 \sum_{k>0} \text{Cov}(X_0, X_k)$ and $n^{-1/2} S_n$ converges in distribution to the normal law $N(0, \Gamma)$.*

Proof For a and b in $\mathbb{R}^d$, denote by $a.b$ the Euclidean scalar product of a and b. Now

$$a.S_n = (a.X_1 - \mathbb{E}(a.X_1)) + \cdots + (a.X_n - \mathbb{E}(a.X_n))$$

for any a in $\mathbb{R}^d$. Consequently, by Theorem 4.2, $n^{-1/2} a.S_n$ converges in distribution to the normal law $N(0, \sigma^2(a))$, with

$$\sigma^2(a) = \text{Var}(a.X_0) + 2 \sum_{k>0} \text{Cov}(a.X_0, a.X_k) = a.\Gamma a.$$

Hence

$$\lim_{n \to \infty} \mathbb{E}(\exp(ia.n^{-1/2} S_n)) = \exp(a.\Gamma a/2).$$

Corollary 4.1 then follows from the Paul Lévy theorem.

4.3 A Functional Central Limit Theorem for the Partial Sum Process

In this section, we give an extension of the functional central limit theorem of Donsker to strongly mixing and stationary sequences of real-valued random variables. We refer to Sects. 9 and 10 in Billingsley (1968) for the definition of the functional central limit theorem and to Sect. 14 in Billingsley (1968) for the weak convergence in the Skorohod space $D([0, 1])$.

Theorem 4.3 below is the functional version of Theorem 4.2. This result improves the functional central limit theorems of Oodaira and Yoshihara (1972) and Herrndorf

(1985), which hold under conditions (IBR) and (HER), respectively. We refer to Merlevède et al. (2006) for a survey of functional central limit theorems for dependent random variables and to Gordin and Peligrad (2011) for functional central limit theorems via martingale approximation.

Theorem 4.3 *Let W denote the usual Wiener measure on $[0, 1]$ (see Billingsley (1968) for a definition) and let $(X_i)_{i \in \mathbb{Z}}$ be a stationary sequence of real-valued and centered random variables satisfying condition (DMR) with the usual strong mixing coefficients, which are defined by (2.1). Let $\{Z_n(t) : t \in [0, 1]\}$ be the normalized partial sum process defined by $Z_n(t) = n^{-1/2} \sum_{i=1}^{[nt]} X_i$, the square brackets designating the integer part.*

Let σ be the nonnegative finite number defined by $\sigma^2 = \lim_n n^{-1} \mathrm{Var} S_n$. Then Z_n converges in distribution to σW in the Skorohod space $D([0, 1])$.

Proof Let Q denote the quantile function of $|X_0|$. We start by proving the finite-dimensional convergence of Z_n to σW. Let $t_0 < t_1 < \ldots < t_k$ be any increasing sequence of reals in $[0, 1]$ such that $t_0 = 0$ and $t_k = 1$. We have to prove that the random vector $(Z_n(t_j) - Z_n(t_{j-1}))_{1 \leq j \leq k}$ converges in distribution to $(\sigma W_{t_j} - \sigma W_{t_{j-1}})_{1 \leq j \leq k}$. Let φ_n denote the characteristic function of $(Z_n(t_j) - Z_n(t_{j-1}))_{1 \leq j \leq k}$, which is defined by

$$\varphi_n(x) = \mathbb{E} \exp\left(i \sum_{j=1}^{k} x_j (Z_n(t_j) - Z_n(t_{j-1}))\right) \text{ for } x = (x_1, x_2, \ldots x_k).$$

Let $\varepsilon > 0$ be any positive real such that $\varepsilon < \inf_{1 \leq j \leq k}(t_j - t_{j-1})$. Let

$$\varphi_{n,\varepsilon}(x) = \mathbb{E} \exp\left(i \sum_{j=1}^{k} x_j (Z_n(t_j - \varepsilon) - Z_n(t_{j-1}))\right).$$

Since the function $y \to e^{iy}$ is 1-Lipschitz,

$$|\varphi_n(x) - \varphi_{n,\varepsilon}(x)| \leq \sum_{j=1}^{k} \|x_j(Z_n(t_j - \varepsilon) - Z_n(t_j))\|_1 \leq 4\|x\|_1 \sqrt{\varepsilon} M_{2,\alpha}(Q) \quad (4.10)$$

by Corollary 1.1(b). Now, by Inequality (6) of Exercise 7, Chap. 1 applied repeatedly k times,

$$\left|\varphi_{n,\varepsilon}(x) - \prod_{j=1}^{k} \mathbb{E} \exp\left(i x_j(Z_n(t_j - \varepsilon) - Z_n(t_{j-1}))\right)\right| \leq 4k\alpha_{[n\varepsilon]-1},$$

which ensures the asymptotic independence of the above increments. Together with Theorem 4.1, this inequality ensures that

$$\lim_{n \to \infty} \varphi_{n,\varepsilon}(x) = \exp\left(\sigma^2 \sum_{j=1}^{k} (t_j - t_{j-1} - \varepsilon) x_j^2\right). \tag{4.11}$$

The finite-dimensional convergence then follows from both (4.10) and (4.11).

It remains to prove the tightness property for the sequence of partial-sum processes $(Z_n)_n$. According to Billingsley (1968), Sect. 8, Theorems 8.2 and 8.4, the tightness property holds in the stationary case if the sequence $(n^{-1} S_n^{*2})_{n>0}$ is uniformly integrable. Hence Proposition 4.1 below completes the proof of Theorem 4.2.

Proposition 4.1 *Let $(X_i)_{i \in \mathbb{Z}}$ be a strictly stationary sequence of centered real-valued random variables satisfying condition (DMR) of Corollary 1.2 for the mixing coefficients defined in (2.30). Set $S_n^* = \sup_{k \in [0,n]} S_k$. Then the sequence $(n^{-1} S_n^{*2})_{n>0}$ is uniformly integrable.*

Proof of Proposition 4.1. Proposition 4.1 is implied by

$$\lim_{A \to +\infty} \sup_{n>0} n^{-1} \mathbb{E}\left((S_n^* - A\sqrt{n})_+^2\right) = 0. \tag{4.12}$$

Now, applying Theorem 3.1(a), we get that

$$n^{-1} \mathbb{E}\left((S_n^* - A\sqrt{n})_+^2\right) \le 16 \int_0^{p_n} \alpha^{-1}(u) Q^2(u) du,$$

with $p_n = \mathbb{P}(S_n^* > A\sqrt{n})$. Now, by the Chebyshev inequality together with Theorem 3.1(b),

$$p_n \le A^{-2} \mathbb{E}(S_n^{*2}/n) \le 16 M_{2,\alpha}(Q) A^{-2}.$$

Hence the above two inequalities imply (4.12), which completes the proof of Proposition 4.1. ∎

4.4 A Central Limit Theorem for Strongly Mixing Triangular Arrays*

In this section, we adapt the Lindeberg (1922) method to strongly mixing sequences. The extensions to mixing sequences started with Bergström (1972) in the stationary ϕ-mixing case (see Krieger (1984) for remarks on Bergström's paper). Dehling (1983) extended the method to strong mixing conditions and random vectors.

Let $(X_{in})_{i\in[1,n]}$ be a triangular array of independent random variables with mean zero and finite variance. Suppose that $\mathrm{Var}(X_{1n} + \cdots + X_{nn}) = 1$. Let $S_{nn} = X_{1n} + \cdots + X_{nn}$. Lindeberg (1922) proved that S_{nn} converges in distribution to a standard normal law as n tends to ∞ if

$$\sum_{i=1}^{n} \mathbb{E}(X_{in}^2 \, \mathbb{I}_{|X_{in}|>\varepsilon}) \to 0 \quad \text{as } n \to \infty \text{ for any } \varepsilon > 0. \tag{4.13}$$

Now one can easily prove that (4.13) holds true if and only if

$$\sum_{i=1}^{n} \mathbb{E}(X_{in}^2 \, \min(|X_{in}|, 1)) \to 0 \quad \text{as } n \to \infty. \tag{4.14}$$

Let U be a random variable with uniform law over $[0, 1]$. Since the random variable $Q_{X_{in}}(U)$ has the same distribution as $|X_{in}|$, (4.14) is equivalent to

$$\sum_{i=1}^{n} \int_{0}^{1} Q_{X_{in}}^2(x) \min(Q_{X_{in}}(x), 1)dx \to 0 \quad \text{as } n \to \infty. \tag{4.15}$$

In this section, we obtain a generalization of the Lindeberg condition to strongly mixing sequences by replacing $Q_{X_{in}}$ by $\alpha^{-1}Q_{X_{in}}$ and dx by $dx/\alpha^{-1}(x)$. Theorem 4.4 below is due to Rio (1995c). We refer to Neumann (2013) for a variant under other types of dependence, with applications to statistics.

Theorem 4.4 *Let m be a positive integer and $(X_{in})_{n\geq m, 1\leq i\leq n}$ be a double array of real-valued random variables with mean zero and finite variance. Let $(\alpha_{k,n})_{k\geq 0}$ be the sequence of strong mixing coefficients in the sense of (2.1) of the sequence $(X_{in})_{i\in[1,n]}$ and $\alpha_{(n)}^{-1}$ be the inverse function of the associated mixing rate function. Let us define $S_{in} = X_{1n} + \cdots + X_{in}$ and $V_{i,n} = \mathrm{Var}\, S_{in}$. Suppose that*

$$V_{n,n} = 1 \quad \text{and} \quad \limsup_{n\to\infty} \max_{i\in[1,n]} V_{i,n} < \infty. \tag{a}$$

Let $Q_{i,n} = Q_{X_{in}}$. Assume furthermore that

$$\lim_{n\to\infty} \left(\sum_{i=1}^{n} \int_{0}^{1} \alpha_{(n)}^{-1}(x) Q_{i,n}^2(x) \min\left(\alpha_{(n)}^{-1}(x) Q_{i,n}(x), 1\right) dx \right) = 0. \tag{b}$$

Then S_{nn} converges in distribution to a standard normal law as n tends to ∞.

Remark 4.2 Theorem 4.2 in the case $\sigma > 0$ follows from Theorem 4.4 applied to $X_{in} = (\mathrm{Var}\, S_n)^{-1/2} X_i$ via Lebesgue's dominated convergence theorem.

Proof of Theorem 4.4. The main step of the proof is the proposition below, which gives quantitative estimates of the accuracy of the characteristic function of a sum

to the characteristic function of a normal law. In order to state this result, we need
some additional notation.

Definition 4.1 For any nonnegative quantile function Q and any positive t, let

$$M_{3,\alpha}(Q, t) = \int_0^1 \alpha^{-1}(x) Q^2(x)(t\alpha^{-1}(x)Q(x) \wedge 1)dx.$$

The proposition below provides an estimate with an error term depending on the
above truncated moments.

Proposition 4.2 *Let $(X_i)_{i\in\mathbb{N}}$ be a strongly mixing sequence of real-valued random
variables with finite variance and mean zero. For any positive k, let $S_k = X_1 +
\cdots + X_k$ and $\varphi_k(t) = \mathbb{E}(\exp(itS_k))$. Set $V_k = \operatorname{Var} S_k$ and $V_n^* = \sup_{k\in[1,n]} V_k$. Let
$Q_k = Q_{X_k}$. For any positive integer n and any real t,*

$$|\exp(V_n t^2/2)\varphi_n(t) - 1| \leq 16t^2 \exp(V_n^* t^2/2) \sum_{k=1}^n M_{3,\alpha}(Q_k, |t|).$$

Before proving Proposition 4.2, we complete the proof of Theorem 4.4. Since

$$M_{3,\alpha}(Q, |t|) \leq \max(|t|, 1)M_{3,\alpha}(Q, 1),$$

Proposition 4.2 ensures that

$$|e^{t^2/2}\varphi_n(t) - 1| \leq 16t^2(|t| \vee 1)e^{V_n^* t^2/2} \sum_{i=1}^n \int_0^1 \alpha_{(n)}^{-1}(x)Q_{i,n}^2(x)\big(\alpha_{(n)}^{-1}(x)Q_{i,n}(x) \wedge 1\big)dx,$$

*with $V_n^* = \max_{i\in[1,n]} V_{i,n}$. Now, under assumption (a), the sequence $(V_n^*)_n$ is a
bounded sequence and, under assumption (b), the above sum converges to 0 as
n tends to ∞. Consequently*

$$\lim_{n\to\infty} |\exp(t^2/2)\varphi_n(t) - 1| = 0,$$

which implies Theorem 4.4 via the Paul Lévy theorem. ∎

Proof of Proposition 4.2. Considering the random variables $-X_i$ if $t < 0$, we may
assume that $t > 0$. Let $V_0 = 0$, $S_0 = 0$ and φ_0 be the characteristic function of S_0.
Set $v_k = V_k - V_{k-1}$ and let

$$\Delta_k = \varphi_k(t) - e^{-v_k t^2/2}\varphi_{k-1}(t). \tag{4.16}$$

Then

$$|e^{t^2/2}\varphi_n(t) - 1| \leq \sum_{k=1}^n e^{V_k t^2/2}|\Delta_k|. \tag{4.17}$$

Let

$$\Delta_{k,1} = \varphi_k(t) - (1 - v_k t^2/2)\varphi_{k-1}(t) \text{ and } \Delta_{k,2} = (1 - v_k t^2/2 - e^{-v_k t^2/2})\varphi_{k-1}(t).$$
(4.18)

Clearly $\Delta_k = \Delta_{k,1} + \Delta_{k,2}$. From (4.17) and the fact that $|\varphi_{k-1}(t)| \leq 1$,

$$|e^{V_n t^2/2}\varphi_n(t) - 1| \leq e^{V_n^* t^2/2}\Big(\sum_{k=1}^n (|\Delta_{k,1}| + g(v_k t^2/2)\Big),$$
(4.19)

with $g(u) = \min(1, e^u)|1 - u - e^{-u}|$. Now, it is easy to check that $g(u) \leq \psi(u)$, with

$$\psi(u) = u^2/2 \text{ for } u \in [-1, 1] \text{ and } \psi(u) = |u| - (1/2) \text{ for } u \notin [-1, 1]. \quad (4.20)$$

Hence

$$|e^{V_n t^2/2}\varphi_n(t) - 1| \leq e^{V_n^* t^2/2}\Big(\sum_{k=1}^n |\Delta_{k,1}| + \sum_{k=1}^n \psi(v_k t^2/2) \Big).$$
(4.21)

Consequently Proposition 4.2 follows from the upper bounds below via (4.21).

Proposition 4.3 *For any positive t,*

$$\sum_{k=1}^n |\Delta_{k,1}| \leq \frac{38}{3}t^2 \sum_{k=1}^n M_{3,\alpha}(Q_k, t)$$
(a)

and

$$\sum_{k=1}^n \psi(v_k t^2/2) \leq \frac{10}{3}t^2 \sum_{k=1}^n M_{3,\alpha}(Q_k, t).$$
(b)

Proof of Proposition 4.3(b). By definition,

$$v_k = \operatorname{Var} X_k + 2\sum_{j=1}^{k-1} \operatorname{Cov}(X_j, X_k).$$

Hence, by Theorem 1.1(a) applied to the random variables X_j and X_k,

$$|v_k| \leq 4\int_0^1 Q_k(x)M_k(x)dx \text{ with } M_k(x) = \sum_{j=1}^k Q_j(x)\, \mathbb{I}_{x<\alpha_{k-j}}.$$
(4.22)

We now introduce some notation.

Definition 4.2 Let $R_k = \alpha^{-1}Q_k$ and let H_k denote the generalized inverse function of R_k. Set $u_k = H_k(1/t)$,

$$x_k = \int_0^{u_k} Q_k(x)M_k(x)dx \ \text{ and } \ y_k = \int_{u_k}^1 Q_k(x)M_k(x)dx.$$

From (4.22) together with the elementary inequality

$$\psi(x+y) \le x + (y^2/2) \tag{4.23}$$

applied to $x = 2x_k t^2$ and $y = 2y_k t^2$, we get that

$$\sum_{k=1}^n \psi(v_k t^2/2) \le 2t^2 \sum_{k=1}^n x_k + 2t^4 \sum_{k=1}^n y_k^2. \tag{4.24}$$

Hence Proposition 4.3(b) follows from the lemma below.

Lemma 4.1 *With the notations of Definitions 4.1 and 4.2,*

$$\sum_{k=1}^n (x_k + t^2 y_k^2) \le \frac{5}{3} \sum_{k=1}^n M_{3,\alpha}(Q_k, t).$$

Proof of Lemma 4.1. By definition of H_k, $t R_k(x) \ge 1$ for any x in $]0, u_k[$. Hence

$$x_k \le \sum_{j=1}^k \int_0^{u_k \wedge \alpha_{k-j}} Q_j(x)Q_k(x)(t R_k(x) \wedge 1)dx. \tag{4.25}$$

Now, from Lemma G.1(a) applied to $a = t R_j(x)$ and $c = t R_k(x)$,

$$Q_j(x)Q_k(x)(t R_k(x) \wedge 1) \le \frac{2}{3}Q_k^2(x)(t R_k(x) \wedge 1) + \frac{1}{2}Q_j^2(x)(t R_j(x) \wedge 1), \tag{4.26}$$

which ensures that

$$x_k \le \sum_{j=1}^k \int_0^{\alpha_{k-j} \wedge u_k} \left(\frac{2}{3}Q_k^2(x)(t R_k(x) \wedge 1) + Q_j^2(x)(t R_j(x) \wedge 1) \right)dx. \tag{4.27}$$

Next, by the Schwarz inequality

$$y_k^2 \le \int_{u_k}^1 Q_k^2(x)M_k^2(x)dx = \sum_{j=1}^k \sum_{l=1}^k \int_{u_k}^1 Q_k^2(x)Q_j(x)Q_l(x) \mathbb{1}_{x<\alpha_{k-j}\wedge\alpha_{k-l}}dx.$$

Now, applying the elementary inequality $2Q_j(x)Q_l(x) \leq Q_j^2(x) + Q_l^2(x)$ and noting that $\sum_{m=0}^{k-1} \mathbb{I}_{x<\alpha_m} \leq \alpha^{-1}(x)$ we obtain that

$$(ty_k)^2 \leq \sum_{j=1}^{k} \int_{u_k}^{1} t^2\alpha^{-1}(x)Q_k^2(x)Q_j^2(x)\,\mathbb{I}_{x<\alpha_{k-j}}\,dx.$$

Now, $\alpha^{-1}(x) \leq (\alpha^{-1}(x))^2$ and, for $x > u_k$, $tR_k(x) \leq 1$. It follows that

$$t^2\alpha^{-1}(x)Q_k^2(x) \leq (tR_k(x))^2 \leq tR_k(x) \wedge 1 \text{ for any } x > u_k.$$

Hence

$$(ty_k)^2 \leq \sum_{j=1}^{k} \int_{u_k}^{1} (tR_k(x) \wedge 1)Q_j^2(x)\,\mathbb{I}_{x<\alpha_{k-j}}\,dx.$$

By Lemma G.1(b) applied to $a = tR_j(x)$ and $c = tR_k(x)$, proceeding as in the proof of (4.26), we get that

$$(tR_k(x) \wedge 1)Q_j^2(x) \leq \frac{1}{3}Q_k^2(x)(tR_k(x) \wedge 1) + Q_j^2(x)(tR_j(x) \wedge 1), \quad (4.28)$$

which ensures that

$$(ty_k)^2 \leq \sum_{j=1}^{k} \int_{\alpha_{k-j}\wedge u_k}^{\alpha_{k-j}} \Big(\frac{2}{3}Q_k^2(x)(tR_k(x) \wedge 1) + Q_j^2(x)(tR_j(x) \wedge 1)\Big)dx. \quad (4.29)$$

Adding (4.27) and (4.29) and summing on k, we then get Lemma 4.1, which completes the proof of Proposition 4.3(b). ∎

Proof of Proposition 4.3(a). Let us give another expression for $\Delta_{1,k}$. Define the function χ_t by $\chi_t(x) = \exp(itx)$. Then $\chi_t'' = -t^2\chi_t$. Hence

$$\Delta_{1,k} = \mathbb{E}\big(\chi_t(S_k)) - \chi_t(S_{k-1}) - \tfrac{1}{2}v_k\chi_t''(S_{k-1})\big). \quad (4.30)$$

Let us now define the class $\mathcal{F}(1, t)$ of regular functions as follows.

Definition 4.3 Let $\mathcal{F}(1, t)$ be the class of real-valued twice differentiable functions f such that $\|f''\|_\infty \leq 1$ and f'' is t-Lipschitz, that is $|f''(x) - f''(y)| \leq t|x - y|$ for any reals x and y.
    Define

$$D_k = \sup_{f\in\mathcal{F}(1,t)} \mathbb{E}\big(f(S_{k-1} + X_k) - f(S_{k-1}) - \tfrac{1}{2}v_k f''(S_{k-1})\big). \quad (4.31)$$

We start by comparing $\Delta_{1,k}$ and D_k.

Lemma 4.2 *For any positive real t, $|\Delta_{1,k}| \leq t^2 D_k$.*

Proof of Lemma 4.2. From the polar decomposition of a complex number, there exists some real θ such that $|\Delta_{1,k}| = \Delta_{1,k} e^{-i\theta}$. It follows that

$$
\begin{aligned}
|\Delta_{1,k}| &= \mathbb{E}\left(e^{i(tS_k-\theta)} - e^{i(tS_{k-1}-\theta)}(1 - \tfrac{1}{2}v_k t^2)\right) \\
&= \mathbb{E}\left(\cos(tS_k - \theta)) - (1 - \tfrac{1}{2}v_k t^2)\cos(tS_{k-1} - \theta)\right). \quad (4.32)
\end{aligned}
$$

For any real θ, the function g_θ defined by $g_\theta(x) = t^{-2}\cos(tx - \theta)$ belongs to $\mathcal{F}(1, t)$. Furthermore, $g_\theta'' = -t^2 g_\theta$. Hence Lemma 4.2 follows from (4.32). ■

From Lemma 4.2, Proposition 4.3(a) follows from the more general upper bound below.

Proposition 4.4 *Under the assumptions of Proposition 4.2,*

$$
\sum_{k=1}^{n} D_k \leq \frac{38}{3} \sum_{k=1}^{n} M_{3,\alpha}(Q_k, t).
$$

Proof of Proposition 4.4. Throughout the proof, we make the convention that $X_i = S_i = 0$ for any $i \leq 0$. The main step of the proof is the following upper bound for D_k.

Lemma 4.3 *Let u be any real in $[0, 1/2]$. Set $\bar{Q}_k(x) = \min(Q_k(x), Q_k(u))$ and $p = \alpha^{-1}(u)$. Then*

$$
\begin{aligned}
D_k \leq & 4 \int_0^u M_k(x) Q_k(x)\,dx \\
&+ 2t \sum_{j=0}^{p-1} \sum_{l=0}^{j+p-1} \int_0^{\alpha_j \wedge \alpha_{(l-j)+}} (1 + \mathbb{1}_{l\in[j,2j-1]}) Q_{k-l}(x) Q_{k-j}(x) \bar{Q}_k(x)\,dx.
\end{aligned}
$$

Proof of Lemma 4.3. Throughout the proof, we make the convention that $S_i = 0$ for any $i \leq 0$. We set

$$
M_k(x, u) = \sum_{j=0}^{p-1} Q_{k-j}(x)\, \mathbb{1}_{x<\alpha_j} \quad \text{and} \quad \bar{X}_k = (X_k \wedge Q_k(u)) \vee (-Q_k(u)). \quad (4.33)
$$

Let $\bar{Q}_k$ denote the quantile function of $|\bar{X}_k|$. From the definition of $\bar{X}_k$,

$$
Q_{\bar{X}_k}(u) = \bar{Q}_k(x) \quad \text{and} \quad Q_{X_k - \bar{X}_k}(x) = (Q_k(x) - Q_k(u))_+ . \quad (4.34)
$$

Let f be any element of $\mathcal{F}(1, t)$. By the Taylor integral formula,

$$f(S_k) - f(S_{k-1}) - f'(S_{k-1})X_k = X_k \int_0^1 (f'(S_{k-1} + vX_k) - f'(S_{k-1}))dv$$

$$= X_k \int_0^1 (f'(S_{k-1} + vX_k) - f'(S_{k-1} + v\bar{X}_k))dv$$

$$+ X_k\bar{X}_k \int_0^1\int_0^1 vf''(S_{k-1} + vv'\bar{X}_k)dvdv'. \quad (4.35)$$

The first term on the right-hand side is bounded above by $|X_k(X_k - \bar{X}_k)|/2$. Moreover,

$$\left| \int_0^1\int_0^1 vf''(S_{k-1} + vv'\bar{X}_k)dvdv' - \frac{1}{2}f''(S_{k-1}) \right| \le \frac{t}{6}|\bar{X}_k|,$$

which ensures that the second term is bounded above by $|X_k\bar{X}_k^2/6|$. Now, from (4.34)

$$\mathbb{E}|X_k(X_k - \bar{X}_k)| = \int_0^u Q_k(x)(Q_k(x) - Q_k(u))dx \quad (4.36)$$

and

$$\mathbb{E}|X_k\bar{X}_k^2| = \int_0^1 Q_k(x)\bar{Q}_k^2(x)dx \le 2\int_0^{1/2} Q_k(x)\bar{Q}_k^2(x)dx. \quad (4.37)$$

Hence

$$\mathbb{E}\left(f(S_k) - f(S_{k-1}) - f'(S_{k-1})X_k - \frac{1}{2}f''(S_{k-1})X_k\bar{X}_k\right) \le$$

$$\frac{1}{2}\int_0^u Q_k(x)(Q_k(x) - Q_k(u))dx + \frac{t}{3}\int_0^{1/2} Q_k(x)\bar{Q}_k^2(x)dx.$$

$$(4.38)$$

We now control the second-order term

$$D_{k,2}(f) = \mathbb{E}\left(f''(S_{k-1})X_k\bar{X}_k\right) - \mathbb{E}(f''(S_{k-1}))\mathbb{E}(X_k\bar{X}_k). \quad (4.39)$$

Let $\Gamma_{k,j} = f''(S_{k-j}) - f''(S_{k-j-1})$. Clearly

$$f''(S_{k-1})X_k\bar{X}_k = \sum_{j=1}^{p-1}\Gamma_{k,j}X_k\bar{X}_k + f''(S_{k-p})X_k\bar{X}_k. \quad (4.40)$$

Since $|\Gamma_{k,j}| \leq t|X_{k-j}|$, applying Theorem 1.1(a) applied to $X = \Gamma_{k,i}$ and $Y = X_k \bar{X}_k$, we get that

$$|\mathrm{Cov}(\Gamma_{k,j}, X_k \bar{X}_k)| \leq 2t \int_0^{\alpha_j} Q_{k-j}(x) Q_k(x) \bar{Q}_k(x) dx. \tag{4.41}$$

Noting that $\alpha_p \leq u$, we also get that

$$|\mathrm{Cov}(f''(S_{k-p}), X_k \bar{X}_k)| \leq 2 \int_0^u Q_k(x) \bar{Q}_k(x) dx. \tag{4.42}$$

The above two inequalities and the decomposition (4.40) together then yield

$$D_{k,2}(f) \leq 2 \int_0^{1/2} \left(t(M_k(x,u) - Q_k(x)) + \mathbb{1}_{x<u} \right) Q_k(x) \bar{Q}_k(x) dx. \tag{4.43}$$

Next, by (4.36)

$$\mathbb{E}(f''(S_{k-1}))\mathbb{E}(X_k(\bar{X}_k - X_k)) \leq \|f''\|_\infty \int_0^u Q_k(x)(Q_k(x) - Q_k(u)) dx. \tag{4.44}$$

Combining (4.38), (4.43) and (4.44) we then get that

$$\mathbb{E}\left(f(S_k) - f(S_{k-1}) - f'\ (S_{k-1})X_k - \frac{1}{2} f''(S_{k-1})\mathbb{E}(X_k^2) \right) \leq$$
$$\int_0^u Q_k^2(x) dx + t \int_0^{1/2} M_k(x,u) Q_k(x) \bar{Q}_k(x) dx. \tag{4.45}$$

It remains to estimate the first-order term $\mathbb{E}(f'(S_{k-1})X_k)$. Let

$$D_{k,1}(f) = \mathbb{E}(f'(S_{k-1})X_k) - \sum_{j=1}^{k-1} \mathbb{E}(f''(S_{k-1}))\mathbb{E}(X_{k-j}X_k). \tag{4.46}$$

In order to bound $D_{k,1}(f)$ we introduce the decomposition below

$$D_{k,1}(f) = \sum_{j=1}^{k-1} D_{k,1}^j(f), \qquad \text{where}$$

$$D_{k,1}^j(f) = \mathrm{Cov}(f'(S_{k-j}) - f'(S_{k-j-1}), X_k) - \mathbb{E}(f''(S_{k-1}))\mathbb{E}(X_{k-j}X_k) \tag{4.47}$$

We now consider two cases. If $j \geq p$, then $\alpha_j \leq u$. Since

$$|f'(S_{k-j}) - f'(S_{k-j-1})| \leq |X_{k-j}|, \tag{4.48}$$

it follows from Theorem 1.1(a) that

$$\text{Cov}(f'(S_{k-j}) - f'(S_{k-j-1}), X_k) \leq 2 \int_0^{\alpha_j} Q_{k-j}(x) Q_k(x) dx. \tag{4.49}$$

Now, by Theorem 1.1(a) applied to $X = X_{k-j}$ and $Y = X_k$,

$$|\mathbb{E}(f''(S_{k-1}))\mathbb{E}(X_{k-j}X_k)| \leq 2\|f''\|_\infty \int_0^{\alpha_j} Q_{k-j}(x) Q_k(x) dx. \tag{4.50}$$

Hence

$$\sum_{j \geq p} D_{k,1}^j(f) \leq 4 \sum_{j \geq p} \int_0^{\alpha_j} Q_{k-j}(x) Q_k(x) dx. \tag{4.51}$$

If $j < p$, we write

$$D_{k,1}^j(f) = \bar{D}_{k,1}^j(f) + \tilde{D}_{k,1}^j(f), \tag{4.52a}$$

with

$$\bar{D}_{k,1}^j(f) = \text{Cov}(f'(S_{k-j}) - f'(S_{k-j-1}), \bar{X}_k) - \mathbb{E}(f''(S_{k-1}))\mathbb{E}(X_{k-j}\bar{X}_k). \tag{4.52b}$$

From the definition of $\tilde{D}_{k,1}^j(f)$ and the fact that $|f''| \leq 1$,

$$\tilde{D}_{k,1}^j(f) \leq |\text{Cov}(f'(S_{k-j}) - f'(S_{k-j-1}), X_k - \bar{X}_k)| + \mathbb{E}|X_{k-j}(X_k - \bar{X}_k)|$$
$$\leq 2 \int_0^{u \wedge \alpha_j} Q_{k-j}(x)(Q_k(x) - \bar{Q}_k(x)) dx + \int_0^u Q_{k-j}(x)(Q_k(x) - \bar{Q}_k(x)) dx$$

by Theorem 1.1(a) together with (4.48) and Lemma 2.1(a). Since $u \wedge \alpha_j \leq u$, we get that

$$\tilde{D}_{k,1}^j(f) \leq 3 \int_0^u Q_{k-j}(x)(Q_k(x) - \bar{Q}_k(x)) dx. \tag{4.53}$$

We now bound $\bar{D}_{k,1}^j(f)$. Let

$$R_{k,j} = f'(S_{k-j}) - f'(S_{k-j-1}) - f''(S_{k-j-1})X_{k-j}. \tag{4.54}$$

By the Taylor formula of order two, $|R_{k,j}| \leq t X_{k-j}^2 / 2$. Consequently, applying Theorem 1.1(a),

$$\text{Cov}(R_{k,j}, \bar{X}_k) \leq t \int_0^{\alpha_j} Q_{k-j}^2(x) \bar{Q}_k(x) dx. \tag{4.55}$$

We now estimate

$$\bar{D}_{k,1}^{j,1}(f) := \bar{D}_{k,1}^j(f) - \text{Cov}(R_{k,j}, \bar{X}_k). \tag{4.56}$$

Here we introduce the decomposition below

$$f''(S_{k-j-1}) = f''(S_{k-2j}) + \sum_{l=1}^{j-1} \Gamma_{k,j+l}. \tag{4.57}$$

Now recall that $|\Gamma_{k,j+l}| \leq t|X_{k-j-l}|$. Hence, by Theorem 1.1(a) applied to $X = \Gamma_{j,k+l}X_{k-j}$ and $Y = \bar{X}_k$ and Lemma 2.1(b),

$$\text{Cov}(\Gamma_{k,j+l}X_{k-j}, \bar{X}_k) \leq 2t \int_0^{\alpha_j} Q_{k-j-l}(x)Q_{k-j}(x)\bar{Q}_k(x)dx. \tag{4.58}$$

We now bound the remainder term

$$\bar{D}_{k,1}^{j,2}(f) := \text{Cov}(f''(S_{k-2j})X_{k-j}, \bar{X}_k) - \mathbb{E}(f''(S_{k-1}))\mathbb{E}(X_{k-j}\bar{X}_k). \tag{4.59}$$

Here we use the decomposition

$$\begin{aligned} \bar{D}_{k,1}^{j,2}(f) = {}& \text{Cov}(f''(S_{k-2j}), X_{k-j}\bar{X}_k) + \mathbb{E}(f''(S_{k-2j})X_{k-j})\mathbb{E}(X_k - \bar{X}_k) \\ &+ \mathbb{E}(f''(S_{k-2j}) - f''(S_{k-1}))\mathbb{E}(X_{k-j}\bar{X}_k). \end{aligned} \tag{4.60}$$

Using Lemma 2.1(a) and noticing that $\alpha_j \geq u$ for $j < p$, we get that

$$|\mathbb{E}(f''(S_{k-2j})X_{k-j})\mathbb{E}(X_k - \bar{X}_k)| \leq \int_0^u Q_{k-j}(x)\,\mathbb{1}_{x<\alpha_j}(Q_k(x) - \bar{Q}_k(x))dx. \tag{4.61}$$

Next

$$|f''(S_{k-2j}) - f''(S_{k-1})| \leq t \sum_{l=1}^{2j-1} |X_{k-l}| \tag{4.62}$$

and, by Theorem 1.1(a) applied to $X = X_{k-j}$ and $Y = \bar{X}_k$,

$$|\mathbb{E}(X_{k-j}\bar{X}_k)| \leq \int_0^{\alpha_j} Q_{k-j}(x)\bar{Q}_k(x)dx, \tag{4.63}$$

whence

$$\mathbb{E}(f''(S_{k-2j}) - f''(S_{k-1}))\mathbb{E}(X_{k-j}\bar{X}_k) \leq 2t \sum_{l=1}^{2j-1} \int_0^{\alpha_j} Q_{k-l}(x)Q_{k-j}(x)\bar{Q}_k(x)dx. \tag{4.64}$$

It remains to bound $\text{Cov}(f''(S_{k-2j}), X_{k-j}\bar{X}_k)$. Clearly

$$f''(S_{k-2j}) = \sum_{l=j}^{p-1} \Gamma_{k,l+j} + f''(S_{k-j-p}). \tag{4.65}$$

Now, by Theorem 1.1(a) applied to $X = \Gamma_{k,l+j}$ and $Y = X_{k-j}\bar{X}_k$ and Lemma 2.1(b),

$$\sum_{l=j}^{p-1} \text{Cov}(\Gamma_{k,l+j}, X_{k-j}\bar{X}_k) \le 2t \sum_{l=j}^{p-1} \int_0^{\alpha_j \wedge \alpha_l} Q_{k-j-l}(x) Q_{k-j}(x) \bar{Q}_k(x) dx. \tag{4.66}$$

Noting that $\alpha_p \le u \le \alpha_j$, and applying Theorem 1.1(a) with $X = f''(S_{k-i-p})$ and $Y = X_{k-i}\bar{X}_k$ together with Lemma 2.1(b), we also get that

$$\text{Cov}(f''(S_{k-j-p}), X_{k-j}\bar{X}_k) \le 2 \int_0^u \mathbb{1}_{x<\alpha_j} Q_{k-j}(x) \bar{Q}_k(x) dx. \tag{4.67}$$

Using the decomposition (4.60), and adding the inequalities (4.61), (4.64), (4.66) and (4.67), we then get that

$$\bar{D}_{k,1}^{j,2}(f) \le \int_0^u Q_{k-j}(x) \mathbb{1}_{x<\alpha_j}(Q_k(x) + \bar{Q}_k(x)) dx$$
$$+ 2t \sum_{l=1}^{j+p-1} \int_0^{\alpha_j \wedge \alpha_{(l-j)+}} Q_{k-l}(x) Q_{k-j}(x) \bar{Q}_k(x) dx.$$

Next, from (4.53), (4.55), (4.58) and the decomposition (4.52), for any j in $[1, p-1]$,

$$D_{k,1}^j(f) \le \int_0^u Q_{k-j}(x) \mathbb{1}_{x<\alpha_j}(4Q_k(x) - 2\bar{Q}_k(x)) dx$$
$$+ 2t \sum_{l=1}^{j+p-1} \int_0^{\alpha_j \wedge \alpha_{(l-j)+}} (1 + \mathbb{1}_{l\in[j,2j-1]}) Q_{k-l}(x) Q_{k-j}(x) \bar{Q}_k(x) dx$$
$$\tag{4.68}$$

Summing (4.68) on j for j in $[1, p-1]$ and adding (4.51), we get that

$$D_{k,1}(f) \le 4 \int_0^u (M_k(x) - Q_k(x)) Q_k(x) dx$$
$$+ 2t \sum_{j=1}^{p-1} \sum_{l=1}^{j+p-1} \int_0^{\alpha_j \wedge \alpha_{(l-j)+}} (1 + \mathbb{1}_{l\in[j,2j-1]}) Q_{k-l}(x) Q_{k-j}(x) \bar{Q}_k(x) dx.$$
$$\tag{4.69}$$

Adding (4.45), we then obtain Lemma 4.3. ∎

End of the proof of Proposition 4.4. Replacing the random variables X_k by $t X_k$ if necessary, we may assume that $t = 1$. Let $u_k = H_k(1)$, $p_k = \alpha^{-1}(u_k)$ and $\bar{Q}_k(x) = \min(Q_k(x), Q_k(u_k))$. Applying Lemma 4.3 with $u = u_k$, we get that

$$\sum_{k=1}^{n} D_k \le 4 \sum_{k=1}^{n} \int_{0}^{u_k} M_k(x) Q_k(x) dx + \sum_{k=1}^{n} I_k \qquad (4.70a)$$

with

$$I_k = \sum_{j=0}^{p_k-1} \sum_{l=0}^{j+p_k-1} \int_{0}^{\alpha_j \wedge \alpha_{(l-j)+}} (1 + \mathbb{1}_{l \in [j, 2j-1]}) 2 Q_{k-l}(x) Q_{k-j}(x) \bar{Q}_k(x) dx. \qquad (4.70b)$$

We now bound the first term on the right-hand side in (4.70a). By definition of u_k, $R_k(x) \ge 1$ for $x < u_k$. It follows that

$$\int_{0}^{u_k} M_k(x) Q_k(x) dx \le \sum_{j=1}^{k} \int_{0}^{\alpha_{k-j}} R_j(x) R_k(x) (R_k(x) \wedge 1) (\alpha^{-1}(x))^{-2} dx. \qquad (4.71)$$

Now, by Lemma G.1(a),

$$R_j(x) R_k(x) (R_k(x) \wedge 1) \le \tfrac{1}{2} R_j^2(x) (R_j(x) \wedge 1) + \tfrac{2}{3} R_k^2(x) (R_k(x) \wedge 1).$$

Putting this inequality in the right-hand side of (4.71) and summing on k, we obtain that

$$\sum_{k=1}^{n} \int_{0}^{u_k} M_k(x) Q_k(x) dx \le \frac{7}{6} \sum_{k=1}^{n} M_{3,\alpha}(Q_k, 1). \qquad (4.72)$$

We now bound I_k. From the inequality $2 Q_{k-l}(x) Q_{k-j}(x) \le Q_{k-l}^2(x) + Q_{k-j}^2(x)$,

$$I_k \le I_k^{(1)} + I_k^{(2)} \text{ with } I_k^{(1)} = \sum_{j=0}^{p_k-1} \sum_{l=0}^{j+p_k-1} \int_{0}^{\alpha_j \wedge \alpha_{(l-j)+}} (1 + \mathbb{1}_{l \in [j, 2j-1]}) Q_{k-j}^2(x) \bar{Q}_k(x) dx$$

$$\text{and } I_k^{(2)} = \sum_{j=0}^{p_k-1} \sum_{l=0}^{j+p_k-1} \int_{0}^{\alpha_j \wedge \alpha_{(l-j)+}} (1 + \mathbb{1}_{l \in [j, 2j-1]}) Q_{k-l}^2(x) \bar{Q}_k(x) dx.$$

$$(4.73)$$

In order to manage $I_k^{(1)}$, we write

$$I_k^{(1)} = \sum_{j=0}^{p_k-1} \int_{0}^{1} n_j(x) Q_{k-j}^2(x) \bar{Q}_k(x) dx \text{ with } n_j(x) = \sum_{l=0}^{j+p_k-1} (1 + \mathbb{1}_{l \in [j, 2j-1]}) \mathbb{1}_{x < \alpha_{(l-j)+} \wedge \alpha_j}.$$

Next

$$n_j(x) = (3j+1)\,\mathbb{1}_{x<\alpha_j} + \sum_{m=j+1}^{p_k-1} \mathbb{1}_{x<\alpha_m} \le 3(\alpha^{-1}(x) \wedge p_k)\,\mathbb{1}_{x<\alpha_j}. \tag{4.74}$$

Since $(\alpha^{-1}(x) \wedge p_k)\bar{Q}_k(x) \le (R_k(x) \wedge 1)$, it follows that

$$I_k^{(1)} \le 3 \sum_{j=0}^{k-1} \int_0^{\alpha_j} Q_{k-j}^2(x)(R_k(x) \wedge 1)\,dx. \tag{4.75}$$

In a similar way

$$I_k^{(2)} \le \sum_{l=0}^{2p_k-2} \int_0^1 N_l(x)Q_{k-l}^2(x)\bar{Q}_k(x)\,dx \quad \text{with } N_l(x) = \sum_{j=0}^{p_k-1}(1 + \mathbb{1}_{l\in[j,2j-1]})\,\mathbb{1}_{x<\alpha_{(l-j)_+} \wedge \alpha_j}.$$

Now $\min(\alpha_{(l-j)_+}, \alpha_j) \le \min(\alpha_{l-[l/2]}, \alpha_j)$. Consequently,

$$N_l(x) \le \mathbb{1}_{x<\alpha_{l-[l/2]}}\left(\sum_{j=0}^{p_k-1} \mathbb{1}_{x<\alpha_j} + \sum_{\substack{j\in]l/2,l] \\ j<p_k}} \mathbb{1}_{x<\alpha_j}\right).$$

If $l < p_k$, then

$$\sum_{\substack{j\in]l/2,l] \\ j<p_k}} \mathbb{1}_{x<\alpha_j} \le \frac{1}{2}\sum_{j\in]l/2,l]}\left(\mathbb{1}_{x<\alpha_j} + \mathbb{1}_{x<\alpha_{l-j}}\right) \le \frac{1}{2}\sum_{m=0}^{p_k-1} \mathbb{1}_{x<\alpha_m}.$$

Otherwise $l \ge p_k$ and

$$\sum_{\substack{j\in]l/2,l] \\ j<p_k}} \mathbb{1}_{x<\alpha_j} \le \sum_{j\in]p_k/2,p_k[} \mathbb{1}_{x<\alpha_j} \le \frac{1}{2}\sum_{m=0}^{p_k-1} \mathbb{1}_{x<\alpha_m}$$

again. From the above inequalities $N_l(x) \le \frac{3}{2}\left(\alpha^{-1}(x) \wedge p_k\right)\mathbb{1}_{x<\alpha_{l-[l/2]}}$, whence

$$I_k^{(2)} \le \frac{3}{2}\sum_{l=0}^{k-1} \int_0^{\alpha_{l-[l/2]}} Q_{k-l}^2(x)(R_k(x) \wedge 1)\,dx. \tag{4.76}$$

Now (4.75) and (4.76) together with (4.28) ensure that

$$\sum_{k=1}^{n} I_k \le 2 \sum_{m=1}^{n} \Big(\sum_{l=0}^{n-1} \int_0^{\alpha_{l-[l/2]}} Q_m^2(x)(R_m(x) \wedge 1)dx + 2 \sum_{j=0}^{n-1} \int_0^{\alpha_j} Q_m^2(x)(R_m(x) \wedge 1)dx \Big).$$

Since $\sum_{l=0}^{n-1} \mathbb{1}_{x<\alpha_{l-[l/2]}} \le 2 \sum_{j=0}^{n-1} \mathbb{1}_{x<\alpha_j} \le 2\alpha^{-1}(x)$, the above inequality implies that

$$\sum_{k=1}^{n} I_k \le 8 \sum_{m=1}^{n} M_{3,\alpha}(Q_m, 1). \tag{4.77}$$

Proposition 4.4 then follows from (4.70), (4.72) and (4.77). ∎

Exercises

(1) **A non-Gaussian limit law.** Let $(\varepsilon_i)_{i\in\mathbb{Z}}$ be a sequence of Gaussian random variables with common distribution $N(0, 1)$ and $V = (a, b)$ be a random variable with values in the unit circle, independent of the sequence $(\varepsilon_i)_{i\in\mathbb{Z}}$. We set $X_i = a\varepsilon_{i-1} + b\varepsilon_i$.

(a) Prove that the coefficients $(\alpha_k)_{k\ge 0}$ defined by (2.30) satisfy $\alpha_k = 0$ for any $k \ge 2$.

(b) Prove that $n^{-1/2} S_n$ converges in distribution to $(a + b)Y$, where Y is an $N(0, 1)$-distributed random variable, independent of $V = (a, b)$. Give a necessary and sufficient condition on V ensuring that the limit law is a Gaussian one.

Problem. - *Agrégation de mathématiques 1994* - Our aim in this problem is to provide a second proof of the central limit theorem for stationary and strongly mixing sequences. We follow the approach of Bolthausen (1982a), which is based on the Stein method (1972). Throughout the problem, $(X_i)_{i\in\mathbb{Z}}$ is a strictly stationary sequence of centered real-valued random variables satisfying condition (DMR) for the strong mixing coefficients $(\alpha_n)_{n\ge 0}$ defined by (2.1). Furthermore, we assume that

$$\sigma^2 = \sum_{i\in\mathbb{Z}} \mathrm{Cov}(X_0, X_i) > 0.$$

A

Let $(\nu_n)_{n>0}$ be a sequence of probability measures on $\mathbb{R}$ such that

$$K := \sup_{n>0} \int_{\mathbb{R}} x^2 d\nu_n(x) < \infty. \tag{0}$$

Suppose furthermore that, for any real λ,

$$\lim_{n \to \infty} \int_{\mathbb{R}} (i\lambda - x) \exp(i\lambda x) d\nu_n(x) = 0. \tag{1}$$

(1) Prove that if $(\nu_n)_{n>0}$ converges weakly to a probability measure ν, then ν is the standard normal law.

(2) Deduce from (1) and from (0) that $(\nu_n)_{n>0}$ converges in distribution to the standard normal law.

B

Throughout part **B**, we assume that $\|X_0\|_\infty = M < \infty$. Let $(m_n)_{n>0}$ be a nondecreasing sequence of positive integers converging to ∞ and such that $m_n \le n/2$ for any $n > 0$. For j in $[1, n]$, we set

$$D_n = \{(l, j) \in [1, n]^2 : |j - l| \le m_n\} \text{ and } D_n(j) = \{l \in [1, n] : |j - l| \le m_n\}.$$

Let

$$V_n = \sum_{(l,j) \in D_n} \mathrm{Cov}(X_j, X_l).$$

(1) Prove that $(V_n/n)_{n>0}$ converges to σ^2 as n tends to ∞.

Throughout the rest of Part **B**, we assume that n is large enough to ensure that $V_n > 0$. We set, for l in $\mathbb{Z}$,

$$Y_{l,n} = V_n^{-1/2} X_l, \quad T_n(j) = \sum_{l \in D_n(j)} Y_{l,n} \text{ and } T_n = \sum_{l=1}^{n} Y_{l,n}.$$

Here λ is any real.

(2) Prove that

$$\mathbb{E}((i\lambda - T_n) \exp(i\lambda T_n)) = i\lambda \mathbb{E}(\exp(i\lambda T_n) A_n) - \mathbb{E}(\exp(i\lambda T_n) B_n) - \mathbb{E}(C_n)$$

with

$$A_n = 1 - \sum_{j=1}^{n} T_n(j) Y_{j,n}, \quad B_n = \sum_{j=1}^{n} Y_{j,n}(1 - \exp(-i\lambda T_n(j)) - i\lambda T_n(j))$$

and

$$C_n = \sum_{j=1}^{n} Y_{j,n} \exp(i\lambda T_n - i\lambda T_n(j)).$$

(3a) Apply the Taylor integral formula to show that

$$|\exp(i\lambda x) - i\lambda x - 1| \leq (\lambda x)^2/2.$$

(3b) Prove that there exists some positive constant K_1 such that

$$\mathbb{E}(|B_n|) \leq K_1 n^{-1/2} m_n$$

for n large enough.

(3c) Prove that there exists some positive constant K_2 such that

$$|\mathbb{E}(C_n)| \leq K_2 n^{1/2} \alpha_{m_n}$$

for n large enough.

(4) Let m be a nonnegative integer and (j, l, j', l') be an element of $\mathbb{Z}^4$ such that $|j - l| \leq m$ and $|j' - l'| \leq m$.

(a) If $|j - j'| \geq 2m$, prove that

$$|\mathrm{Cov}(X_j X_l, X_{j'} X_{l'})| \leq 2M^4 \alpha_{|j-j'|-2m}.$$

(b) If $k = \min(|j - j'|, |j - l|, |j - l'|)$, prove that

$$|\mathrm{Cov}(X_j X_l, X_{j'} X_{l'})| \leq 4M^4 \alpha_k.$$

(5) Prove that $\mathbb{E}(A_n) = 0$. Next, prove that there exists some positive constant K_3 such that

$$\mathbb{E}(A_n^2) \leq K_3 n^{-1} m_n^2$$

for n large enough.

(6a) Prove that the sequence $(m\alpha_m)_{m>0}$ converges to 0. Find a sequence $(m_n)_{n>0}$ of positive integers with the above prescribed properties such that

$$\lim_{n\to\infty} n^{1/2} \alpha_{m_n} = \lim_{n\to\infty} n^{-1/2} m_n = 0.$$

(6b) Prove then that $n^{-1/2} S_n$ converges in distribution to the law $N(0, \sigma^2)$.

C

Let M be any positive real. We set

$$f_M(x) = x \, \mathbb{1}_{|x| \leq M}.$$

We denote by H the tail function defined by $H(x) = \mathbb{P}(|X_0| > x)$ and by Q the càdlàg inverse of H. Let

$$Z_n = n^{-1/2} \sum_{j=1}^{n} X_j, \quad \bar{Z}_{n,M} = n^{-1/2} \sum_{j=1}^{n} (f_M(X_j) - \mathbb{E}(f_M(X_j)))$$

and

$$\tilde{Z}_{n,M} = Z_n - \bar{Z}_{n,M}.$$

(1) Prove that

$$\mathbb{E}(\tilde{Z}_{n,M}^2) \le 4 \int_0^{H(M)} \alpha^{-1}(u) Q^2(u) du.$$

(2a) Prove that the series

$$\sum_{k \in \mathbb{Z}} \mathrm{Cov}(f_M(X_0), f_M(X_k))$$

is absolutely convergent.

(2b) Let $\sigma^2(M)$ be the sum of the above series. Prove that $\sigma^2(M)$ converges to σ^2 as M tends to ∞.

(2c) Prove that the central limit theorem holds under condition (DMR).

(2) *A central limit theorem for β-mixing sequences.* Let $(X_i)_{i \in \mathbb{Z}}$ be a strictly stationary sequence of random variables with values in some Polish space $\mathcal{X}$, with common law P. We assume that the sequence of strong mixing coefficients defined in (2.1) is summable and that the sequence of β-mixing coefficients $(\beta_i)_{i \ge 0}$ defined in Corollary 1.4 is summable. Let B be defined as in Corollary 1.4 and $Q = BP$.

(a) Prove that, for any g in $L^2(Q)$, the series

$$\sum \mathrm{Cov}(g(X_0), g(X_t))$$

is absolutely convergent. Bound the sum $\sigma^2(g)$ of this series.

(b) Proceed as in part **C** of the problem to prove that

$$Z_n(g) = n^{-1/2}(S_n(g) - \mathbb{E}(S_n(g)))$$

converges in distribution to the law $N(0, \sigma^2(g))$.

Chapter 5
Mixing and Coupling

5.1 Introduction

One of the most popular techniques to obtain limit theorems for dependent processes is to replace the initial sequence by a sequence with finite range dependence. In this direction, the coupling lemmas allow one to replace the initial sequence after time 0 by a new sequence independent of the past before time 0. In this chapter, we give coupling theorems for mixing sequences. The cost of the coupling will depend on the mixing condition involved. Here we will give coupling results for strongly mixing or absolutely regular sequences.

For sequences of random variables satisfying a β-mixing condition, the new sequence is equal to the initial sequence after time n with high probability. This result was obtained independently by Berbee (1979) and Goldstein (1979). This result fails in the strong mixing case. Nevertheless, one can still obtain weaker results which are efficient for real-valued random variables. These results are stated and proved in Sect. 5.2 in the case of bounded random variables. Next, in Sect. 5.3, we will state and prove coupling lemmas for random variables satisfying a β-mixing condition. In Sect. 5.4 we compare the results of Sect. 5.2 to previous results on the same subject. In Sect. 5.5, we give the strong version of Berbee's or Goldstein's Lemma, called maximal coupling. Section 5.6 is devoted to an extension of the results of Scct. 5.2 to unbounded random variables.

5.2 A Coupling Lemma for Real-Valued Random Variables

We first state the coupling lemma of Berbee (1979) for random variables satisfying a β-mixing condition. This lemma will be proved in Sect. 5.3.

Lemma 5.1 *Let A be a σ-field in $(\Omega, \mathcal{T}, \mathbb{P})$ and X be a random variable with values in some Polish space $\mathcal{X}$. Let δ be a random variable with uniform distribution over*

© Springer-Verlag GmbH Germany 2017
E. Rio, *Asymptotic Theory of Weakly Dependent Random Processes*,
Probability Theory and Stochastic Modelling 80,
DOI 10.1007/978-3-662-54323-8_5

[0, 1], *independent of the σ-field generated by X and* $\mathcal{A}$. *Then there exists a random variable* X^*, *with the same law as X, independent of X, such that* $\mathbb{P}(X \neq X^*) = \beta(\mathcal{A}, \sigma(X))$. *Furthermore,* X^* *is measurable with respect to the σ-field generated by* $\mathcal{A}$ *and* (X, δ).

When X is a real-valued random variable with values in the compact interval $[a, b]$, Lemma 5.1 ensures that

$$\mathbb{E}(|X - X^*|) \leq (b - a)\beta(\mathcal{A}, \sigma(X)). \tag{5.1}$$

In this section we will prove that (5.1) is still true if one replaces the β-mixing coefficient by the strong mixing coefficient, and more generally, by the dependence coefficient defined in (1.8b).

Lemma 5.2 *Let* $\mathcal{A}$ *be a σ-field in* $(\Omega, \mathcal{T}, \mathbb{P})$ *and X be a real-valued random variable with values in* $[a, b]$. *Let* δ *be a random variable with uniform distribution over* $[0, 1]$, *independent of the σ-field generated by X and* $\mathcal{A}$. *Then there exists a random variable* X^*, *with the same law as X, independent of X, such that*

$$\mathbb{E}(|X - X^*|) \leq (b - a)\alpha(\mathcal{A}, X).$$

Furthermore, X^* *is measurable with respect to the σ-field generated by* $\mathcal{A}$ *and* (X, δ).

Proof We will define X^* from X via the conditional quantile transformation. The main interest of the quantile transformation is that this transformation minimizes the L^1-distance between X and X^*. We refer to Major (1978) for the properties of the quantile transformations.

Let F be the distribution function of X, and $F_{\mathcal{A}}$ be the conditional distribution function of X given $\mathcal{A}$, which is defined by $F_{\mathcal{A}}(t) = \mathbb{P}(X \leq t \mid \mathcal{A})$. Since δ is independent of $\mathcal{A} \vee \sigma(X)$ and has the uniform distribution over $[0, 1]$, the random variable

$$V = F_{\mathcal{A}}(X - 0) + \delta(F_{\mathcal{A}}(X) - F_{\mathcal{A}}(X - 0)) \tag{5.2}$$

has the uniform distribution over $[0, 1]$, conditionally to $\mathcal{A}$ (see Annex F). Hence V is independent of $\mathcal{A}$ and has the uniform distribution over $[0, 1]$. Therefore

$$X^* = F^{-1}(V) \tag{5.3}$$

is independent of $\mathcal{A}$ and has the same distribution function as X. Furthermore (see Exercise 1, Chap. 1),

$$X = F_{\mathcal{A}}^{-1}(V) \quad \text{a.s.,} \tag{5.4}$$

whence

$$\mathbb{E}(|X - X^*|) = \mathbb{E}\left(\int_0^1 |F_{\mathcal{A}}^{-1}(v) - F^{-1}(v)| dv\right). \tag{5.5}$$

Since X takes its values in $[a, b]$,

$$\int_0^1 |F_{\mathcal{A}}^{-1}(v) - F^{-1}(v)| dv = \int_a^b |F_{\mathcal{A}}(t) - F(t)| dt.$$

Interchanging the integral and the mean, we infer that

$$\mathbb{E}(|X - X^*|) = \int_a^b \mathbb{E}(|F_{\mathcal{A}}(t) - F(t)|) dt. \tag{5.6}$$

Now, by (1.10c), for any real t,

$$\mathbb{E}(|F_{\mathcal{A}}(t) - F(t)|) \leq \alpha(\mathcal{A}, X), \tag{5.7}$$

which, together with (5.6) implies Lemma 5.2.   ∎

5.3   A Coupling Lemma for β-Mixing Random Variables

In this section, we give a constructive proof of Lemma 5.1. In Sect. 5.4 we will study connections between coupling for β-mixing random variables and coupling for strongly mixing random variables.

Proof of Lemma 5.1 Let X^* be a random variable, independent of $\mathcal{A}$ and with the same distribution as X. For any pair $(A_i)_{i \in I}$ and $(B_j)_{j \in J}$ of finite partitions of Ω and $\mathcal{X}$, with A_i in $\mathcal{A}$ and B_j a Borel subset of $\mathcal{X}$,

$$\sum_{i \in I} \sum_{j \in J} |\mathrm{Cov}(\mathbb{1}_{A_i}, \mathbb{1}_{X \in B_j})| = \sum_{i \in I} \sum_{j \in J} |\mathbb{P}(A_i \cap (X \in B_j)) - \mathbb{P}(A_i \cap (X^* \in B_j))|$$

$$\leq \sum_{i \in I} \mathbb{E}\left(\mathbb{1}_{A_i} \sum_{j \in J} |\mathbb{1}_{X \in B_j} - \mathbb{1}_{X^* \in B_j}|\right).$$

Now $\sum_{j \in J} |\mathbb{1}_{X \in B_j} - \mathbb{1}_{X^* \in B_j}| \leq 2\mathbb{1}_{X \neq X^*}$, and consequently

$$\frac{1}{2} \sum_{i \in I} \sum_{j \in J} |\mathrm{Cov}(\mathbb{1}_{A_i}, \mathbb{1}_{X \in B_j})| \leq \mathbb{P}(X \neq X^*). \tag{5.8}$$

Therefore, by (1.58), $\mathbb{P}(X \neq X^*) \geq \beta(\mathcal{A}, \sigma(X))$.

Let us now prove the converse inequality. From Lemma E.1 in Annex E, it is enough to prove Lemma 5.1 for random variables X with values in $([0, 1], \mathcal{B})$, where $\mathcal{B}$ denotes the σ-field of Borel sets of $[0, 1]$. We start with the construction of the random variables in $(\Omega \times [0, 1] \times [0, 1], \mathcal{A} \otimes \mathcal{B} \otimes \mathcal{B})$.

Here, we use the notation introduced in the proof of Lemma 5.2. We have to construct a probability measure on the above product space in such a way that, if Y denotes the second canonical projection and Y^* denotes the third canonical projection, then

$$\mathbb{P}(Y \leq t \mid \mathcal{A}) = F_{\mathcal{A}}(t), \ \mathbb{P}(Y^* \leq t \mid \mathcal{A}) = F(t) \text{ and } \mathbb{P}(Y \neq Y^*) \leq \beta. \quad (5.9)$$

On the first component we consider the probability induced on $\mathcal{A}$ by $\mathbb{P}$. In order to define the law on the product space, it is enough to define the conditional law $\nu_{\mathcal{A}}$ of (Y, Y^*) conditionally to ω.

Notation 5.1 For L a nonnegative integer, let $I_{L,1} = [0, 2^{-L}]$ and $I_{L,i} =]$ $(i - 1)2^{-L}, i2^{-L}]$ for i in $[2, 2^L]$. Let $\mathcal{B}_L$ be the Boolean algebra generated by the sets $I_{L,i}$.

We now define a coherent sequence $(\nu_{L,\mathcal{A}})_L$ of conditional probabilities on the algebras $\mathcal{B}_L \otimes \mathcal{B}_L$. The conditional probability $\nu_{\mathcal{A}}$ will be defined from these conditional probabilities via an extension theorem.

Assume that a coherent sequence $(\nu_{L,\mathcal{A}})_{L \leq N}$ of conditional laws on the Boolean algebras $\mathcal{B}_L \otimes \mathcal{B}_L$ has been constructed in such a way that these laws are measurable with respect to $\mathcal{A}$ and satisfy the condition $\mathcal{H}(L)$ below: if $p_{i,j}^L = \nu_{L,\mathcal{A}}(I_{L,i} \times I_{L,j})$, then, for any L in $[0, N]$ and any i in $[1, 2^L]$,

$$p_{i,i}^L = \mathbb{P}(X \in I_{L,i} \mid \mathcal{A}) \wedge \mathbb{P}(X \in I_{L,i}), \ \sum_{j=1}^{2^L} p_{i,j}^L = \mathbb{P}(X \in I_{L,i} \mid \mathcal{A}), \ \sum_{i=1}^{2^L} p_{i,j}^L = \mathbb{P}(X \in I_{L,i})$$

(note that $\mathcal{H}(0)$ is satisfied anyway). We have to construct an extension $\nu_{N+1,\mathcal{A}}$ of $\nu_{N,\mathcal{A}}$ to the Boolean algebra $\mathcal{B}_{N+1} \otimes \mathcal{B}_{N+1}$ in such a way that $\mathcal{H}(N + 1)$ still holds true.

For any pair (i, j) of integers in $[1, 2^N]$, the extension $\nu_{N+1,\mathcal{A}}$ has to satisfy the constraints

$$p_{i,j}^N = \sum_{\varepsilon=0}^{1} \sum_{\eta=0}^{1} p_{2i-\varepsilon, 2j-\eta}^{N+1}. \quad \mathcal{C}(N + 1)$$

Furthermore, we need to construct $\nu_{N+1,\mathcal{A}}$ in such a way that $\mathcal{H}(N + 1)$ holds true. Set

$$a_\varepsilon^i = \mathbb{P}(X \in I_{N+1,2i-\varepsilon} \mid \mathcal{A}), \ b_\varepsilon^j = \mathbb{P}(X \in I_{N+1,2j-\varepsilon}) \text{ and } q_{\varepsilon\eta}^{ij} = p_{2i-\varepsilon, 2j-\eta}^{N+1}.$$

$$(5.10)$$

We start by defining the diagonal terms. In order to fulfill $\mathcal{H}(N+1)$, we set

$$q_{\varepsilon\varepsilon}^{ii} = \mathbb{P}(X \in I_{N+1,2i-\varepsilon} \mid \mathcal{A}) \wedge \mathbb{P}(X \in I_{N+1,2i-\varepsilon}). \tag{5.11}$$

Now we have to fulfill both the constraint $\mathcal{C}(N+1)$ for $j = i$ and (5.11). With the notation introduced in (5.10), this means that

$$q_{00}^{ii} = a_0^i \wedge b_0^i, \; q_{11}^{ii} = a_1^i \wedge b_1^i \text{ and}$$
$$q_{01}^{ii} + q_{10}^{ii} = \inf(a_0^i + a_1^i, b_0^i + b_1^i) - (a_0^i \wedge b_0^i) - (a_1^i \wedge b_1^i). \tag{5.12}$$

If $a_0^i + a_1^i \le b_0^i + b_1^i$, then the constraint on the first marginal at range N, which may be written as $\sum_j p_{i,j}^N = a_0^i + a_1^i$, implies that $p_{i,j}^N = 0$ for $j \ne i$, whence $q_{\varepsilon\eta}^{ij} = 0$ for $j \ne i$. Consequently, the first marginal constraints at range $N+1$ for the lines $2i$ and $2i-1$ hold true if and only if

$$q_{01}^{ii} = a_0^i - (a_0^i \wedge b_0^i) \text{ and } q_{10}^{ii} = a_1^i - (a_1^i \wedge b_1^i). \tag{i}$$

Hence there exists a unique $(q_{\varepsilon\eta}^{ii})_{\varepsilon,\eta}$ satisfying (5.12) and the marginal constraints on the lines. Similarly, if $a_0^i + a_1^i > b_0^i + b_1^i$, then the constraint on the first marginal at range N implies that $p_{j,i}^N = 0$ for any $j \ne i$. Then the marginal constraints at range $N+1$ on columns $2i$ and $2i-1$ hold true if and only if

$$q_{10}^{ii} = b_0^i - (a_0^i \wedge b_0^i) \text{ and } q_{01}^{ii} = b_1^i - (a_1^i \wedge b_1^i). \tag{ii}$$

Hence there exists a unique $(q_{\varepsilon\eta}^{ii})_{\varepsilon,\eta}$ satisfying (5.12) and the marginal constraints on the columns.

It remains to define the probabilities $p_{2i-\varepsilon,2j-\eta}^{N+1}$ for $j \ne i$. If $p_{i,j}^N = 0$, then these numbers are equal to 0. If $p_{i,j}^N \ne 0$, then

$$\mathbb{P}(X \in I_{N,i} \mid \mathcal{A}) > \mathbb{P}(X \in I_{N,i}) \text{ and } \mathbb{P}(X \in I_{N,j} \mid \mathcal{A}) < \mathbb{P}(X \in I_{N,j}). \tag{5.13}$$

Under (5.13) the reals q_{10}^{ii} and q_{01}^{ii} are determined by (ii). Summing on the lines $2i$ and $2i-1$, we then get that the marginal constraints are satisfied if

$$\begin{cases} q_{00}^{ii} + q_{01}^{ii} = b_1^i + (a_0^i \wedge b_0^i) - (a_1^i \wedge b_1^i) \le a_0^i \\ q_{10}^{ii} + q_{11}^{ii} = b_0^i + (a_1^i \wedge b_1^i) - (a_0^i \wedge b_0^i) \le a_1^i. \end{cases}$$

Now $b_0^i + b_1^i < a_0^i + a_1^i$, whence $b_0^i < a_0^i$ or $b_1^i < a_1^i$. If $b_0^i < a_0^i$, then

$$\begin{cases} b_1^i + (a_0^i \wedge b_0^i) \le b_0^i + b_1^i \le \inf(a_0^i + a_1^i, a_0^i + b_1^i) \\ b_0^i + (a_1^i \wedge b_1^i) - (a_0^i \wedge b_0^i) \le (a_1^i \wedge b_1^i) \le a_1^i, \end{cases}$$

which ensures that the above marginal constraints are satisfied. The case $b_1^i < a_1^i$ can be treated in a similar way. Hence $r_{i0} = (a_0^i - q_{00}^{ii} - q_{01}^{ii})/(a_0^i + a_1^i - b_0^i - b_1^i)$ is nonnegative. In a similar way, it can be proven that $r_{i0} \le 1$. Let $r_{i1} = 1 - r_{i0}$. We now deal with column j. By (4.13), the reals $q_{\varepsilon\eta}^{jj}$ satisfy condition (i) (with j instead of i). Starting from (i), one can define nonnegative reals s_{j0} and $s_{j1} = 1 - s_{j0}$ corresponding to column j, in the same way as the reals r_{i0} and r_{i1} are defined from (ii). We then set

$$p_{2i-\varepsilon,2j-\eta}^{N+1} = r_{i\varepsilon} s_{j\eta} p_{i,j}^N \text{ for } (\varepsilon, \eta) \in \{0, 1\}^2, \tag{5.14}$$

which completes the definition of $\nu_{N+1,\mathcal{A}}$. Condition $\mathcal{C}(N + 1)$ is then satisfied. We now check the constraint on the first margin (the constraint on the second margin can be checked in a similar way). Under (i) this constraint holds immediately since $p_{ij}^N = 0$ for $j \ne i$. Under (ii), we have to prove that

$$\sum_{j=1}^{2^N} \sum_{\eta=0}^{1} p_{2i,2j-\eta} = a_0^i. \tag{5.15}$$

Now, separating $j = i$ and $j \ne i$ in this sum and using (ii), we get that (5.15) holds true if and only if $r_{i0} \sum_{j \ne i} p_{i,j}^N = (a_0^i - q_{00}^{ii} - q_{01}^{ii})$. Now the constraint at range N on the line i for $\nu_{N,\mathcal{A}}$ may be written as $\sum_{j \ne i} p_{i,j}^N = (a_0^i + a_1^i - b_0^i - b_1^i)$, so that (5.15) holds true by definition of r_{i0}. Consequently, there exists a sequence $(\nu_{N,\mathcal{A}})_N$ of conditional probabilities with the prescribed properties. Define then the probability measure ν_N on $\mathcal{A} \otimes \mathcal{B}_N \otimes \mathcal{B}_N$ by $\nu_N(A \times B_N) = \mathbb{E}(\nu_{N,\mathcal{A}}(B_N) \mathbb{I}_A)$. The so-defined sequence $(\nu_N)_N$ of probability measures is coherent. Hence, by the Kolmogorov extension theorem, there exists a unique probability measure ν on $\mathcal{A} \otimes \mathcal{B} \otimes \mathcal{B}$ such that

$$\nu(A \times B_N) = \nu_N(A \times B_N) = \mathbb{E}(\mathbb{I}_A \nu_{N,\mathcal{A}}(B_N)) \tag{5.16}$$

for any A in $\mathcal{A}$, any positive integer N and any B_N in $\mathcal{B}_N \otimes \mathcal{B}_N$. Let $\nu_{\mathcal{A}}$ be defined by

$$\nu(A \times B) = \mathbb{E}(\mathbb{I}_A \nu_{\mathcal{A}}(B)) \text{ for } A \in \mathcal{A} \text{ and } B \in \mathcal{B} \otimes \mathcal{B}. \tag{5.17}$$

The restriction of $\nu_{\mathcal{A}}$ to $\mathcal{B}_N \otimes \mathcal{B}_N$ is equal to $\nu_{N,\mathcal{A}}$. Hence, for any dyadic number x,

$$\nu_{\mathcal{A}}([0, x] \times [0, 1]) = F_{\mathcal{A}}(x) \text{ and } \nu_{\mathcal{A}}([0, 1] \times [0, x]) = F(x). \tag{5.18}$$

Since the dyadic numbers are dense in $[0, 1]$, it follows that (5.18) holds for any real x in $[0, 1]$. Let Y denote the second canonical projection and Y^* the third canonical projection. From (5.18) the random variable Y^* is independent of $\mathcal{A}$ (here $\mathcal{A}$ denotes the σ-field induced by the first projection) and has the same law as X. From (5.18) again the conditional law of Y given $\mathcal{A}$ is equal to the conditional law of X given $\mathcal{A}$. Furthermore,

$$\mathbb{P}(Y = Y^*) = \lim_{N \to \infty} \mathbb{E}\left(\nu_{N,\mathcal{A}}\left(\bigcup_{i=1}^{2^N} I_{N,i} \times I_{N,i}\right)\right) = \lim_{N \to \infty} \sum_{i=1}^{2^N} \mathbb{E}(p_{i,i}^N). \quad (5.19)$$

Now

$$\sum_{i=1}^{2^N} \mathbb{E}(p_{i,i}^N) = \frac{1}{2} \sum_{i=1}^{2^N} \mathbb{E}(|\mathbb{P}(X \in I_{N,i} \mid \mathcal{A}) - \mathbb{P}(X \in I_{L,i})|) \geq 1 - \beta(\mathcal{A}, \sigma(X)),$$

which ensures that $\mathbb{P}(Y \neq Y^*) \leq \beta(\mathcal{A}, \sigma(X))$. Hence, $\mathbb{P}(Y \neq Y^*) = \beta(\mathcal{A}, \sigma(X))$.

Let $\tilde{\Omega} = \Omega \times [0, 1] \times [0, 1] \times [0, 1]$ be equipped with $\nu \otimes \lambda$. By Lemma E.2, there exists a random variable V with uniform law over $[0, 1]$ independent of the σ-field $\mathcal{G}$ induced by the first two projections and a measurable function g such that $Y^* = g(\omega, Y, V)$ almost surely. Setting $X^* = g(\omega, X, \delta)$, we then get Lemma 5.1.   ■

5.4   Comparison of α-Mixing and β-Mixing Coefficients for Finite σ-Fields

In this section, we are interested in a converse inequality for β-mixing and strong mixing coefficients in the case of σ-fields with a finite number of atoms. Below we give a result of Bradley (1983) which was used to obtain approximation theorems in the case of real-valued random variables. At the end of this section we will compare this lemma with Lemma 5.1.

Lemma 5.3 *Let $\mathcal{A}$ be a σ-field of $(\Omega, \mathcal{T}, \mathbb{P})$ and $\mathcal{B}$ be a Boolean algebra included in $\mathcal{T}$, having exactly K atoms. Then $\beta(\mathcal{A}, \mathcal{B}) \leq (2K)^{1/2}\alpha(\mathcal{A}, \mathcal{B})$.*

Remark 5.1 The above lemma is optimal up to some multiplicative constant, as proved by Bradley (1983). This fact will be proved in Exercise 2.

Proof of Lemma 5.3 We may assume that the probability space is large enough to contain a finite sequence $(\varepsilon_1, \ldots, \varepsilon_K)$ of independent and symmetric signs, independent of $\mathcal{A} \vee \mathcal{B}$. Let $B_1, \ldots, B_K$ denote the atoms of $\mathcal{B}$. Set

$$Y = \sum_{k=1}^{K} \varepsilon_i (\mathbb{1}_{B_k} - \mathbb{P}(B_k)).$$

We now proceed conditionally to $(\varepsilon_1, \ldots, \varepsilon_K)$: the random variable Y is conditionally centered, so that we may apply (1.11c) with $X = 1$ conditionally to the values of the signs. Since Y belongs to $[-1, 1]$, integrating with respect to the signs, we get that

$$\mathbb{E}\left(\left|\sum_{k=1}^{K} \varepsilon_i (\mathbb{P}(B_k \mid \mathcal{A}) - \mathbb{P}(B_k))\right|\right) \le 2\alpha(\mathcal{A}, \mathcal{B}). \tag{5.20}$$

Now, by the lower bound of Szarek (1976) in Khinchin's inequality, for any finite sequence $a_1, \ldots, a_K$ of reals,

$$\mathbb{E}(|a_1\varepsilon_1 + \cdots + a_K\varepsilon_K|) \ge 2^{-1/2}(a_1^2 + \cdots + a_K^2)$$
$$\ge (2K)^{-1/2}(|a_1| + \cdots + |a_K|) \tag{5.21}$$

by the Cauchy–Schwarz inequality. Taking $a_k = \mathbb{P}(B_k \mid \mathcal{A}) - \mathbb{P}(B_k)$ in (5.21), we get that

$$2\alpha(\mathcal{A}, \mathcal{B}) \ge (2K)^{-1/2} \sum_{i=1}^{K} \mathbb{E}(|\mathbb{P}(B_k \mid \mathcal{A}) - \mathbb{P}(B_k)|) = (2/K)^{1/2}\beta(\mathcal{A}, \mathcal{B}),$$

which completes the proof of Lemma 5.3. ∎

We now explain the method of Bradley (1983) for strongly mixing sequences. Divide $[a, b]$ into K intervals $H_1, H_2, \ldots H_K$ of the same length. For X a random variable with values in $[a, b]$, consider the Boolean algebra $\mathcal{B}$ generated by the atoms $B_k = (X \in H_k)$. Applying Lemma 5.1 together with Lemma 5.3, one can construct a random variable X^* (which depends on the number K of intervals) with the same distribution as X, independent of $\mathcal{A}$ and such that

$$\mathbb{P}\left((X, X^*) \in \bigcup_{k=1}^{K} H_k \times H_k\right) \ge 1 - (2K)^{1/2}\alpha(\mathcal{A}, \sigma(X)). \tag{5.22}$$

Now, if (X, X^*) belongs to $H_k \times H_k$ for some k in $[1, K]$, then $|X - X^*| \le (b - a)/K$. Consequently,

$$\mathbb{P}(|X - X^*| > K^{-1}(b - a)) \le (2K)^{1/2}\alpha(\mathcal{A}, \sigma(X)). \tag{5.23}$$

Now, for any λ in $[0, b - a]$, applying (5.23) with $K = 1 + [(b - a)/\lambda]$, we get that

$$\mathbb{P}(|X - X^*| > \lambda) \le 2((b - a)/\lambda)^{1/2}\alpha(\mathcal{A}, \sigma(X)) \tag{5.24}$$

(see Bradley (1983), Theorem 3). The main default of (5.24) is that X^* depends on λ: therefore this inequality cannot be integrated with respect to λ, which leads to a loss for $\mathbb{E}(|X - X^*|)$. From (5.24),

$$\mathbb{E}(|X - X^*|) \le \lambda + (b - a)\mathbb{P}(|X - X^*| > \lambda) \le \lambda + 2\alpha(\mathcal{A}, \sigma(X))(b - a)^{3/2}\lambda^{-1/2}.$$

For the optimal choice $\lambda = (b - a)(\alpha(\mathcal{A}, \sigma(X)))^{2/3}$, the above inequality gives

$$\mathbb{E}(|X - X^*|) \leq 3(b - a)^{1/3}(\alpha(\mathcal{A}, \sigma(X)))^{2/3}. \tag{5.25}$$

For the pair (X, X^*) constructed in the proof of Lemma 5.1, this upper bound can be improved. We refer to Exercise 1 for an upper bound on $\mathbb{E}(|X - X^*|)$ for the pair (X, X^*) constructed in the proof of Lemma 5.1 similar to the upper bound of Lemma 5.2.

5.5 Maximal Coupling and Absolutely Regular Sequences

In this section, we give a relation between maximal coupling and absolutely regular sequences. Theorem 5.1 below, which can be found in Goldstein (1979) and Berbee (1979), generalizes a result of Griffeath (1975) for Markov chains.

Theorem 5.1 *Let $(\xi_i)_{i \in \mathbb{Z}}$ be a sequence of random variables with values in some Polish space $\mathcal{X}$. Assume that $(\Omega, \mathcal{T}, \mathbb{P})$ is rich enough to contain a random variable U with uniform distribution over $[0, 1]$, independent of $(\xi_i)_{i \in \mathbb{Z}}$. Let $\mathcal{F}_0 = \sigma(\xi_i : i \leq 0)$ and $\mathcal{G}_n = \sigma(\xi_i : i \geq n)$. Then one can construct a sequence $(\xi_i^*)_{i \in \mathbb{Z}}$ with the same joint distribution as the initial sequence $(\xi_i)_{i \in \mathbb{Z}}$, independent of $\mathcal{F}_0$ and measurable with respect to the σ-field generated by U and $(\xi_i)_{i \in \mathbb{Z}}$, in such a way that, for any positive integer n,*

$$\mathbb{P}(\xi_k \neq \xi_k^* \text{ for some } k \geq n \mid \mathcal{F}_0) = \text{ess sup } \{|\mathbb{P}(B \mid \mathcal{F}_0) - \mathbb{P}(B)| : B \in \mathcal{G}_n\}. \quad (a)$$

In particular,

$$\mathbb{P}(\xi_k \neq \xi_k^* \text{ for some } k \geq n) = \beta(\mathcal{F}_0, \mathcal{G}_n). \quad (b)$$

Remark 5.2 The β-mixing coefficients of the sequence $(\xi_i)_{i \in \mathbb{Z}}$ are determined uniquely by property (b). Hence Theorem 5.1 contains all the information needed to explore the properties of β-mixing sequences. In Chap. 8 we will apply this result to uniform limit theorems for empirical processes.

5.6 An Extension of Lemma 5.2 to Unbounded Random Variables *

In this section, we give an extension of Lemma 5.2 to unbounded real-valued random variables. The result below is due to Peligrad (2002).

Lemma 5.4 *Let $\mathcal{A}$ be a σ-field in $(\Omega, \mathcal{T}, \mathbb{P})$ and X be a real-valued and integrable random variable. Let δ be a random variable with uniform distribution over*

[0, 1], *independent of the σ-field generated by X and A. Then there exists a random variable X*, with the same law as X, independent of X, such that*

$$\mathbb{E}(|X - X^*|) \leq 2 \int_0^{\alpha(\mathcal{A}, X)} Q_X(u) du.$$

Furthermore, X* is measurable with respect to the σ-field generated by A and (X, δ).

Proof As in Lemma 5.2, the random variable X^* is defined from X by (5.2) and (5.3). For the so-defined random variable X^*, (5.5) still holds true. Proceeding exactly as in the proof of (5.6), we then get that

$$\|X - X^*\|_1 = \int_{\mathbb{R}} \mathbb{E}(|F_{\mathcal{A}}(t) - F(t)|) dt. \tag{5.26}$$

Since $|F_{\mathcal{A}}(t) - F(t)| = |\mathbb{P}(X > t \mid \mathcal{A}) - \mathbb{P}(X > t)|$,

$$\int_0^\infty \mathbb{E}(|F_{\mathcal{A}}(t) - F(t)|) dt = \int_0^\infty \mathbb{E}(|\mathbb{P}(X > t \mid \mathcal{A}) - \mathbb{P}(X > t)|) dt.$$

Next

$$\int_{-\infty}^0 \mathbb{E}(|F_{\mathcal{A}}(t) - F(t)|) dt = \int_{-\infty}^0 \mathbb{E}(|\mathbb{P}(X < t \mid \mathcal{A}) - \mathbb{P}(X < t)|) dt,$$

whence

$$\|X - X^*\|_1 = \int_0^\infty \mathbb{E}(|\mathbb{P}(X > t \mid \mathcal{A}) - \mathbb{P}(X > t)|$$
$$+ |\mathbb{P}(-X > t \mid \mathcal{A}) - \mathbb{P}(-X > t)|) dt. \tag{5.27}$$

Now, by (1.10c),

$$\mathbb{E}(|\mathbb{P}(X > t \mid \mathcal{A}) - \mathbb{P}(X > t)| + |\mathbb{P}(-X > t \mid \mathcal{A}) - \mathbb{P}(-X > t)|) \leq 2\alpha(\mathcal{A}, X).$$

Furthermore,

$$|\mathbb{P}(X > t \mid \mathcal{A}) - \mathbb{P}(X > t)| + |\mathbb{P}(-X > t \mid \mathcal{A}) - \mathbb{P}(-X > t)|$$
$$\leq \mathbb{P}(|X| > t) + \mathbb{P}(|X| > t \mid \mathcal{A}),$$

whence

$$\mathbb{E}(|\mathbb{P}(X > t \mid \mathcal{A}) - \mathbb{P}(X > t)| + |\mathbb{P}(-X > t \mid \mathcal{A}) - \mathbb{P}(-X > t)|) \leq 2\mathbb{P}(|X| > t).$$

Combining (5.27) with the above two upper bounds, we get that

$$\|X - X^*\|_1 \leq 2 \int_0^\infty \inf(\alpha(\mathcal{A}, X), \mathbb{P}(|X| > t))dt.$$

Since

$$\int_0^\infty \inf(\alpha(\mathcal{A}, X), \mathbb{P}(|X| > t))dt = \int_0^{\alpha(\mathcal{A},X)} Q_X(u)du,$$

Lemma 5.4 follows. ∎

Exercises

(1) Let $\mathcal{A}$ be a σ-field of $(\Omega, \mathcal{T}, \mathbb{P})$ and X be a random variable with values $[0, 1]$. Set $\alpha = \alpha(\mathcal{A}, \sigma(X))$ and let X^* be the random variable constructed in the proof of Lemma 5.2.

(a) Prove that, for any positive integer N,

$$\mathbb{P}(\text{there exists an } i \in [1, 2^N] \text{ such that } (X, X^*) \in I_{N,i} \times I_{N,i}) \geq 1 - 2^{(N+1)/2}\alpha.$$

Infer that, for any positive λ, $\mathbb{P}(|X - X^*| > \lambda) \leq 2\lambda^{-1/2}\alpha$.

(b) Prove that $\mathbb{E}(|X - X^*|) \leq 4\alpha$.

We now assume that the random variable X takes its values in $[0, 1]^d$ equipped with the distance d_∞.

(c) Prove that there exists a bijective and bimeasurable transformation from $[0, 1]$ to $[0, 1]^d$ such that, for any positive integer N, the images of the dyadic intervals $I_{N,i}$ are dyadic boxes with diameter bounded above by $2^{-\lfloor N/d \rfloor}$.

(d) Construct a random variable X^* independent of $\mathcal{A}$ and with the same law as X in such a way that

$$\mathbb{P}(d_\infty(X, X^*) > 4\lambda^{1/d}) \leq (2/\lambda)^{1/2}\alpha.$$

From the above upper bound, deduce an upper bound on $\mathbb{E}(d_\infty(X, X^*))$. Compare this upper bound with the upper bound which can be deduced from Lemma 5.1 in the β-mixing case.

(2) **On the optimality of Lemma** 5.3. - Bradley (1983) - Recall that the correlation between two square-integrable random variables X and Y is defined by

$$\text{Corr}(X, Y) = (\text{Var } X \text{Var } Y)^{-1/2}\text{Cov}(X, Y).$$

If $\mathcal{A}$ and $\mathcal{B}$ are two σ-fields in some probability space, we set

$$\rho(\mathcal{A}, \mathcal{B}) = \sup\{\text{Corr}(X, Y) : X \in L^2(\mathcal{A}), Y \in L^2(\mathcal{B})\}.$$

Let N be an even natural integer. Let $\Omega_1 = [0, 1]$ be equipped with its Borel field, which is denoted here by $\mathcal{F}_1$, and P_1 be the Lebesgue measure on $\mathcal{F}_1$. Let $\Omega_2 = \{1, \ldots, N\}$ be equipped with $\mathcal{F}_2 = \mathcal{P}(\Omega_2)$. On Ω_2 we consider the uniform distribution, which is denoted by P_2.

Set $m = N/2$. Let $h_1, h_2, \ldots$ be the Rademacher functions given by $h_j(x) = (-1)^{\lfloor x 2^j \rfloor}$. On $\Omega = \Omega_1 \times \Omega_2$ equipped with $\mathcal{F}_1 \otimes \mathcal{F}_2$, we define the probability measure P as follows: the density with respect to the Lebesgue measure of the conditional law of ω_1 conditionally to $(\omega_2 = j)$ is equal to $1 - h_j(x)$ for j in $[1, m]$ and to $1 - h_{j-m}(x)$ for j in $[m + 1, N]$.

Let $\mathcal{A} = \{F_1 \times \Omega_2 : F_1 \in \mathcal{F}_1\}$ and $\mathcal{B} = \{\Omega_1 \times F_2 : F_2 \in \mathcal{F}_2\}$.

(a) Prove that $\beta(\mathcal{A}, \mathcal{B}) = 1/2$.

(b) Prove that any numerical function g on $\{1, 2, \ldots, N\}$ has the decomposition $g = g_1 + g_2$ with $g_1(j + m) = -g_1(j + m)$ and $g_2(j + m) = g_2(m)$ for any j in $[1, m]$. Prove that this decomposition is unique. Next, prove that, under the law P_2, Var $g \geq$ Var g_1.

(c) Let f be a square integrable Borel function on $[0, 1]$, with mean 0. Prove that $\mathrm{Cov}(f, g) = \mathrm{Cov}(f, g_1)$. Infer that $|\mathrm{Corr}(f, g)| \leq |\mathrm{Corr}(f, g_1)|$.

(d) Let $c_j = g_1(j)$. Prove that

$$\mathrm{Cov}(f, g_1) = \frac{2}{N} \int_0^1 \sum_{j=1}^m c_j h_j(x) f(x) dx.$$

Infer that $\rho(\mathcal{A}, \mathcal{B}) \leq (2/N)^{1/2}$.

(e) Prove that $\rho(\mathcal{A}, \mathcal{B}) \geq 2\alpha(\mathcal{A}, \mathcal{B})$. Conclude that $\alpha(\mathcal{A}, \mathcal{B})(N/2)^{1/2} \leq \beta(\mathcal{A}, \mathcal{B})$.

Chapter 6
Fuk–Nagaev Inequalities, Applications

6.1 Introduction

In this chapter, we generalize the classical exponential inequalities for sums of independent random variables (we refer to Appendix B for these inequalities) to sums of strongly mixing random variables. Our approach is based on coupling, as in Bradley (1983) and Bosq (1993). We improve their results by using Lemma 5.2, which provides a more efficient coupling for strongly mixing random variables. Starting from the initial sequence and applying this coupling lemma, we will replace the initial sequence by a q-dependent sequence of random variables. The cost of this coupling depends on q. We refer to Theorem 2 in Berkes and Philipp (1979) for a similar method in the ϕ-mixing case. Next, applying the usual exponential inequalities for sums of independent random variables to this new sequence, we obtain inequalities with two parts in the upper bound: an exponential term and a term depending mainly on the mixing coefficient α_q. For power-type rates of mixing the second term does not decrease exponentially. This is the reason why our inequalities are similar to the inequalities of Fuk and Nagaev (1971) for sums of unbounded random variables. In Sect. 6.3, we derive a Fuk–Nagaev type inequality for unbounded random variables from the inequalities of Sect. 6.2. Next, in Sect. 6.4, we apply this inequality to get moment inequalities in the style of Rosenthal (1970) and Marcinkiewicz–Zygmund type inequalities. Our method is similar to the method used in Petrov (1989). In Sect. 6.5 we give an application of our Fuk–Nagaev type inequality to the bounded law of the iterated logarithm.

6.2 Exponential Inequalities for Partial Sums

In this section, we apply Lemma 5.2 together with the Bennett inequality for sums of independent random variables to get a new maximal inequality for partial sums of bounded random variables in the strong mixing case.

© Springer-Verlag GmbH Germany 2017
E. Rio, *Asymptotic Theory of Weakly Dependent Random Processes*,
Probability Theory and Stochastic Modelling 80,
DOI 10.1007/978-3-662-54323-8_6

Theorem 6.1 *Let $(X_i)_{i>0}$ be a sequence of real-valued random variables such that $\|X_i\|_\infty \leq M$ for any positive i, and $(\alpha_n)_{n\geq 0}$ be the sequence of strong mixing coefficients defined by (2.1). Set $X_i = 0$ for any $i > n$. Let $S_k = \sum_{i=1}^{k}(X_i - \mathbb{E}(X_i))$. Let q be any positive integer, and v_q be any positive real such that*

$$v_q \geq \sum_{i>0} \mathbb{E}((X_{iq-q+1} + \cdots + X_{iq})^2).$$

Set $M(n) = \sum_{i=1}^{n} \|X_i\|_\infty$ and let $h(x) = (1+x)\log(1+x) - x$. Then, for any $\lambda \geq qM$,

$$\mathbb{P}\left(\sup_{k\in[1,n]} |S_k| \geq (\mathbb{1}_{q>1} + 5/2)\lambda\right) \leq 4\exp\left(-\frac{v_q}{(qM)^2}h\left(\frac{\lambda q M}{v_q}\right)\right) + 4M(n)\frac{\alpha_{q+1}}{\lambda}$$

$$\leq 4\exp\left(-\frac{\lambda}{2qM}\log\left(1 + \frac{\lambda q M}{v_q}\right)\right) + 4M(n)\frac{\alpha_{q+1}}{\lambda}.$$

Proof Set $U_i = S_{iq} - S_{iq-q}$. Since $X_i = 0$ for any $i > n$, the random variables U_i are almost surely equal to 0 for i large enough. Now, for any integer $j, d(j, q\mathbb{Z}) \leq [q/2]$. It follows that

$$\sup_{k\in[1,n]} |S_k| \leq 2[q/2]M + \sup_{j>0} |\sum_{i=1}^{j} U_i|.$$

Hence Theorem 6.1 is a by-product of the inequality below:

$$\mathbb{P}\left(\sup_{j>0} |\sum_{i=1}^{j} U_i| \geq 5\lambda/2\right) \leq 4\exp\left(-\frac{v_q}{(qM)^2}h(\lambda q M/v_q)\right) + 4M(n)\frac{\alpha_{q+1}}{\lambda}. \quad (6.1)$$

The inequality in Theorem 6.1 follows immediately from both (6.1) and the lower bound

$$h(x) \geq x \int_0^1 \log(1+tx)dt \geq x\log(1+x) \int_0^1 tdt \geq x\log(1+x)/2.$$

Proof of Inequality (6.1). Let $(\delta_j)_{j>0}$ be a sequence of independent random variables with uniform law over $[0, 1]$, independent of the sequence $(U_i)_{i>0}$. Applying Lemma 5.2 recursively, we obtain that, for any integer $i \geq 3$, there exists a measurable function F_i such that the random variable $U_i^* = F_i(U_1, ..., U_{i-2}, U_i, \delta_i)$ satisfies the conclusions of Lemma 5.2 with $\mathcal{A} = \sigma(U_l : l < i - 1)$. Set $U_i^* = U_i$ for $i = 1$ and $i = 2$. The so-constructed sequence $(U_i^*)_{i>0}$ has the properties below:

1. For any positive i, the random variable U_i^* has the same distribution as U_i.
2. The random variables $(U_{2i}^*)_{i>0}$ are independent and the random variables $(U_{2i-1}^*)_{i>0}$ are independent.
3. For any integer $i \geq 3$,

$$\mathbb{E}(|U_i - U_i^*|) \leq 2\alpha_{q+1} \sum_{k=iq-q+1}^{iq} \|X_k\|_\infty.$$

Replacing the initial random variables U_i by the random variables U_i^*, we get that

$$\sup_{j>0} |\sum_{i=1}^{j} U_i| \leq \sum_{i\geq 3} |U_i - U_i^*| + \sup_{j>0} |\sum_{i=1}^{j} U_{2i}^*| + \sup_{j>0} |\sum_{i=1}^{j} U_{2i-1}^*|. \tag{6.2}$$

By property 3 together with the Markov inequality,

$$\mathbb{P}\left(\sum_{i>0} |U_i - U_i^*| \geq \lambda/2\right) \leq 4M(n)\alpha_{q+1}\lambda^{-1}. \tag{6.3}$$

To complete the proof of Inequality (6.1), it then suffices to apply Theorem B.1(b) in Appendix B twice with $K = Mq$ and $v = v_q$ to the random variables $(U_{2i}^*)_{i>0}$ and the random variables $(U_{2i-1}^*)_{i>0}$. ∎

6.3   Fuk–Nagaev Inequalities for Partial Sums

In this section, we give an extension of the Fuk–Nagaev inequality for sequences of independent random variables to strongly mixing sequences of random variables. However, in order to get an efficient inequality, we have to assume that the tails of the random variables are uniformly bounded. We refer to Dedecker and Prieur (2004) for an extension of this inequality to a weaker notion of dependence and to Merlevède et al. (2011) for more efficient inequalities in the case of exponential or semi-exponential rates of mixing.

Theorem 6.2 *Let $(X_i)_{i>0}$ be a sequence of real-valued and centered random variables with finite variances. Let $(\alpha_n)_{n\geq 0}$ denote the sequence of strong mixing coefficients defined in (2.1). Set $Q = \sup_{i>0} Q_i$ and*

$$s_n^2 = \sum_{i=1}^{n} \sum_{j=1}^{n} |\text{Cov}(X_i, X_j)|. \tag{6.4}$$

Let $R(u) = \alpha^{-1}(u)Q(u)$ and let $H(u) = R^{-1}(u)$ denote the generalized inverse function of R. Then, for any positive λ and any $r \geq 1$,

$$\mathbb{P}\left(\sup_{k\in[1,n]} |S_k| \geq 4\lambda\right) \leq 4\left(1 + \frac{\lambda^2}{rs_n^2}\right)^{-r/2} + 4n\lambda^{-1} \int_0^{H(\lambda/r)} Q(u)du. \tag{6.5}$$

Remark 6.1 As in Theorem 6.1, we may assume that $X_i = 0$ for $i > n$. Consequently (6.5) remains true if $\alpha^{-1}(u)$ is replaced by $\alpha^{-1}(u) \wedge n$.

Proof We may assume that $X_i = 0$ for any $i > n$. Let q be any positive integer and let M be a positive real. Set

$$U_i = S_{iq} - S_{iq-q} \text{ and } \bar{U}_i = (U_i \wedge qM) \vee (-qM) \text{ for any } i \in \{1, \ldots, [n/q]\}. \quad (6.6)$$

From the assumption $X_i = 0$ for $i > n$, $\bar{U}_i = 0$ for $i > [n/q]$. Let $\varphi_M(x) = (|x| - M)_+$. We start by proving that

$$\sup_{k \in [1,n]} |S_k| \le \sup_{j>0} \left| \sum_{i=1}^{j} \bar{U}_j \right| + qM + \sum_{k=1}^{n} \varphi_M(X_k). \quad (6.7)$$

To prove (6.7), we notice that, if the maximum of the random variables $|S_k|$ is obtained for k_0, then, for $j_0 = [k_0/q]$,

$$\sup_{k \in [1,n]} |S_k| \le \left| \sum_{i=1}^{j_0} \bar{U}_i \right| + \sum_{i=1}^{j_0} |U_i - \bar{U}_i| + \sum_{k=qj_0+1}^{k_0} |X_k|. \quad (6.8)$$

Now, by convexity of the function φ_M,

$$\sum_{i=1}^{j_0} |U_i - \bar{U}_i| \le \sum_{k=1}^{qj_0} \varphi_M(X_k). \quad (6.9)$$

Moreover,

$$\sum_{k=qj_0+1}^{k_0} |X_k| \le (k_0 - qj_0)M + \sum_{k=qj_0+1}^{k_0} \varphi_M(X_k), \quad (6.10)$$

and combining the above three inequalities, we get (6.7).

In order to apply Theorem 6.1, we have to center the random variables $\bar{U}_i$. Since the random variables U_i are centered,

$$\sup_{j>0} \left| \sum_{i=1}^{j} \bar{U}_i \right| \le \sup_{j>0} \left| \sum_{i=1}^{j} (\bar{U}_i - \mathbb{E}(\bar{U}_i)) \right| + \sum_{i>0} \mathbb{E}(|U_i - \bar{U}_i|)$$

$$\le \sup_{j>0} \left| \sum_{i=1}^{j} (\bar{U}_i - \mathbb{E}(\bar{U}_i)) \right| + \sum_{k=1}^{n} \mathbb{E}(\varphi_M(X_k)), \quad (6.11)$$

by convexity of φ_M. Hence,

$$\sup_{k\in[1,n]} |S_k| \le \sup_{j>0} \left| \sum_{i=1}^{j} (\bar{U}_i - \mathbb{E}(\bar{U}_i)) \right| + qM + \sum_{k=1}^{n} (\mathbb{E}(\varphi_M(X_k)) + \varphi_M(X_k)).$$

(6.12)

We now choose M and q. Let $x = \lambda/r$ and $v = H(x)$. If $v = 1/2$, then

$$4n\lambda^{-1} \int_0^{H(\lambda/r)} Q(u)du \ge 2n\lambda^{-1} \int_0^1 Q(u)du \ge 2\lambda^{-1} \sum_{i=1}^{n} \mathbb{E}(|X_i|). \qquad (6.13)$$

In that case, Inequality (6.5) follows immediately from the Markov inequality applied to the random variable $|X_1| + \cdots + |X_n|$. If $v < 1/2$, then $\alpha^{-1}(v) > 0$. In that case, we set

$$q = \alpha^{-1}(v) \ \text{and} \ M = Q(v). \qquad (6.14)$$

With this choice of q and v,

$$qM = R(v) = R(H(x)) \le x \le \lambda. \qquad (6.15)$$

We now apply Theorem 6.1 to the random variables $\bar{U}_i$. Setting $q = 1$ and $M = x$ in Theorem 6.1 and noticing that

$$\mathbb{E}(\bar{U}_i^2) \le \mathbb{E}(U_i^2) \le \sum_{l,m\in]iq-q,iq]} |\mathrm{Cov}(X_l, X_m)|, \qquad (6.16)$$

which ensures that Theorem 6.1 holds with $v_1 = s_n^2$, we get that

$$\mathbb{P}\left(\sup_{j>0} \left| \sum_{i=1}^{j} (\bar{U}_j - \mathbb{E}(\bar{U}_i)) \right| \ge 5\lambda/2 \right) \le 4\left(1 + \frac{\lambda^2}{rs_n^2} \right)^{-r/2} + 4nM\alpha_{q+1}\lambda^{-1}. \quad (6.17)$$

It remains to bound the deviation of the second random variable on the right-hand side of (6.12). By the Markov inequality,

$$\mathbb{P}\left(\sum_{k=1}^{n} (\mathbb{E}(\varphi_M(X_k)) + \varphi_M(X_k)) \ge \lambda/2 \right) \le \frac{4}{\lambda} \sum_{k=1}^{n} \int_0^1 (Q_k(u) - Q(v))_+ du$$

$$\le \frac{4n}{\lambda} \int_0^v (Q(u) - Q(v))du.$$

(6.18)

Since $q \ge \alpha^{-1}(v)$, one can prove that $\alpha_q \le v$ and $M\alpha_{q+1} \le vQ(v)$. Putting together (6.13), (6.17), (6.18), and noting that $Mq \le \lambda$, we then obtain Theorem 6.2. ∎

An application to power-type rates of mixing. Let $(X_i)_{i>0}$ be a strongly mixing sequence. Assume that the strong mixing coefficients α_n satisfy $\alpha_n \le cn^{-a}$ for some constants $c \ge 1$ and $a > 1$. Suppose furthermore that there exists some $p > 2$ such that

$$\mathbb{P}(|X_i| > t) \leq t^{-p} \text{ for any } t > 0.$$

Then, setting $b = ap/(a + p)$, an elementary calculation yields $H(x) \leq c^{b/a}(2/x)^b$, whence

$$4\lambda^{-1} \int_0^{H(\lambda/r)} Q(u)du \leq 4Cr^{-1}(\lambda/r)^{-(a+1)p/(a+p)},$$

with $C = 2p(2p - 1)^{-1}(2^a c)^{(p-1)/(a+p)}$. Consequently, by Theorem 6.2, for any $r \geq 1$ and any positive λ,

$$\mathbb{P}\left(\sup_{k \in [1,n]} |S_k| \geq 4\lambda \right) \leq 4\left(1 + \frac{\lambda^2}{rs_n^2}\right)^{-r/2} + 4Cnr^{-1}(r/\lambda)^{(a+1)p/(a+p)}. \tag{6.19a}$$

If $\|X_i\|_\infty \leq 1$ (which corresponds to $p = \infty$), Theorem 6.2 applied with $Q = 1$ yields

$$\mathbb{P}\left(\sup_{k \in [1,n]} |S_k| \geq 4\lambda \right) \leq 4\left(1 + \frac{\lambda^2}{rs_n^2}\right)^{-r/2} + 2ncr^{-1}(2r/\lambda)^{a+1}. \tag{6.19b}$$

6.4 Application to Moment Inequalities

In this section we adapt to the strong mixing case the method proposed by Petrov (1989) in the independent case to derive moment inequalities in the style of Rosenthal (1970) from the Fuk–Nagaev inequality. Our first result is an extension of Theorem 2.2 for algebraic moments to moments of any order $p > 2$.

Theorem 6.3 *Let $(X_i)_{i>0}$ be a sequence of real-valued and centered random variables and $(\alpha_n)_{n\geq 0}$ be the sequence of strong mixing coefficients defined by (2.1). Suppose that, for some $p > 2$, $\mathbb{E}(|X_i|^p) < \infty$ for any positive integer i. Then*

$$\mathbb{E}\left(\sup_{k \in [1,n]} |S_k|^p \right) \leq a_p s_n^p + nb_p \int_0^1 [\alpha^{-1}(u) \wedge n]^{p-1} Q^p(u)du,$$

where

$$Q = \sup_{i>0} Q_i, \quad a_p = p4^{p+1}(p + 1)^{p/2} \text{ and } b_p = \frac{p}{p - 1}4^{p+1}(p + 1)^{p-1}.$$

Remark 6.2 We refer to Appendix C for more about the quantities involved in these moment inequalities. Note that Q can be replaced by $Q_{(n)} = \sup_{i \in [1,n]} Q_i$ in Theorem 6.3.

Proof As in the proof of Theorems 6.1 and 6.2 we may assume that $X_i = 0$ for any $i > n$. Under this assumption $\alpha^{-1}(u) \leq n$. First,

$$\mathbb{E}\left(\sup_{k\in[1,n]}|S_k|^p\right)= p4^p \int_0^\infty \lambda^{p-1}\mathbb{P}\left(\sup_{k\in[1,n]}|S_k| \geq 4\lambda\right)d\lambda.$$

Now, applying Theorem 6.2 with $r = p + 1$, we get that

$$\mathbb{E}\left(\sup_{k\in[1,n]}|S_k|^p\right)\leq p4^p(4E_2 + 4nE_1), \qquad (6.20)$$

with

$$E_2 = \int_0^\infty \left(1 + \frac{\lambda^2}{rs_n^2}\right)^{-r/2}\lambda^{p-1}d\lambda \quad \text{and} \quad E_1 = \int_0^\infty \int_0^1 \lambda^{p-2}Q(u)\,\mathbb{1}_{u<H(\lambda/r)}d\lambda du.$$

We now bound E_2. Since H is the right-continuous inverse of R, $(H(\lambda/r) > u)$ if and only if $(\lambda < rR(u))$. Hence, interchanging the integrals, we obtain that

$$E_1 \leq \frac{1}{p-1}(p+1)^{p-1}\int_0^1 Q(u)R^{p-1}(u)du. \qquad (6.21)$$

To bound E_2, we introduce the change of variable $x = \lambda/(s_n\sqrt{r})$. Then

$$E_2 = (p+1)^{p/2}s_n^p \int_0^\infty x^{p-2}(1+x^2)^{-(p+1)/2}xdx.$$

Since $x^{p-2} \leq (1+x^2)^{(p-2)/2}$, it follows that

$$E_2 \leq (p+1)^{p/2}s_n^{p/2}\int_0^\infty (1+x^2)^{-3/2}xdx.$$

Consequently,

$$E_2 \leq (p+1)^{p/2}s_n^{p/2}.$$

Both (6.20), (6.21) and the above inequality then imply Theorem 6.3. ∎

Let

$$M_{p,\alpha}(Q) = \int_0^1 [\alpha^{-1}(u)]^{p-1}Q^p(u)du \quad \text{and} \quad M_{p,\alpha,n}(Q) = \int_0^1 [\alpha^{-1}(u)\wedge n]^{p-1}Q^p(u)du.$$

If $M_{p,\alpha}(Q) < \infty$, then Theorem 6.3 yields a Rosenthal type inequality. Since $M_{p,\alpha,n}(Q)$ converges to $M_{p,\alpha}(Q)$ as n tends to infinity, this is not the case if $M_{p,\alpha}(Q) = \infty$. Nevertheless, one can still obtain a Marcinkiewicz–Zygmund type inequality. In order to state this inequality, we need to introduce weak norms.

Definition 6.1 For any real $r \geq 1$ and any real-valued random variable X, we set

$$\Lambda_r(X) = \sup_{t>0}\left(t^r \mathbb{P}(|X| > t)\right)^{1/r}.$$

With this definition $\lim_{r\to\infty} \Lambda_r(X) = \|X\|_\infty$.

Corollary 6.1 below gives a moment inequality which improves on the results of Yokoyama (1980).

Corollary 6.1 *Let $p > 2$ and $(X_i)_{i>0}$ be a sequence of real-valued and centered random variables and $(\alpha_n)_{n\geq 0}$ be the sequence of strong mixing coefficients defined by (2.1). Suppose that, for some $r > p$, $\Lambda_r(X_k) < \infty$ for any positive integer k and that the strong mixing coefficients satisfy*

$$\alpha_n \leq \kappa(n+1)^{-pr/(2r-2p)} \quad \text{for any } n \geq 0, \text{ for some positive } \kappa.$$

Then there exists some positive constant $C(\kappa, p)$ such that

$$\mathbb{E}\left(\sup_{k\in[1,n]} |S_k|^p\right) \leq \frac{r}{r-p}C(\kappa, p)\kappa^{-p/r}n^{p/2}\left(\sup_{k>0}\Lambda_r(X_k)\right)^p.$$

Remark 6.3 Corollary 6.1 still holds in the case $r = \infty$. In that case the mixing coefficients have to satisfy $\alpha_n \leq \kappa(n+1)^{-p/2}$ and

$$\mathbb{E}\left(\sup_{k\in[1,n]} |S_k|^p\right) \leq C(\kappa, p)n^{p/2}\sup_{k>0}\|X_k\|_\infty^p.$$

Proof of Corollary 6.1. Let $K = \sup_{k>0} \Lambda_r(X_k)$. By the Markov inequality,

$$\mathbb{P}(|X_k| > t) \leq (K/t)^r \quad \text{for any positive } t,$$

whence $Q(u) \leq Ku^{-1/r}$ for any u in $[0, 1]$. Now both the above bound on Q and (C.10) ensure that

$$M_{p,\alpha,n}(Q) \leq K^p(p-1)\sum_{i=0}^{n-1}(i+1)^{p-2}\int_0^{\alpha_i} u^{-p/r}du$$

$$\leq K^p\frac{r}{r-p}\kappa^{1-p/r}(p-1)\sum_{j=1}^{n} j^{-2+(p/2)}.$$

Now, for $p \leq 4$,

$$\sum_{j=1}^{n} j^{-2+(p/2)} \leq \int_0^n x^{-2+(p/2)}dx = 2(p-2)^{-1}n^{-1+(p/2)},$$

and, for $p > 4$,

$$\sum_{j=1}^{n} j^{-2+(p/2)} \leq n^{-2+(p/2)} + \int_{1}^{n} x^{-2+(p/2)} dx \leq 2n^{-1+(p/2)}.$$

It follows that

$$n M_{p,\alpha,n}(Q) \leq \frac{r}{r-p} \cdot \frac{2(p-1)}{(p-2) \wedge 1} K^p \kappa^{1-p/r} n^{p/2}. \tag{6.22}$$

We now bound s_n^p. By Corollary 1.1 together with the fact that $Q^2(u) \leq K^2 u^{-2/r}$, we have:

$$s_n^2 \leq 4n K^2 \sum_{i=0}^{n-1} \int_{0}^{\alpha_i} u^{-2/r} du \leq \frac{4nr}{r-2} K^2 \kappa^{1-2/r} \sum_{j=1}^{n} j^{-p(r-2)/(2r-2p)} \leq \frac{4np}{p-2} K^2 \kappa^{1-2/r}$$

under the mixing assumption of Corollary 6.1. Now, the above bound, (6.22) and Theorem 6.3 together imply Corollary 6.1 with

$$C(\kappa, p) = \left(\frac{4p}{p-2}\right)^{p/2} a_p \kappa^{p/2} + \frac{2(p-1)}{(p-2) \wedge 1} b_p \kappa.$$

6.5 Application to the Bounded Law of the Iterated Logarithm

The first known results on the law of the iterated logarithm for strongly mixing sequences are due to Oodaira and Yoshihara (1971a, b). Later Rio (1995b) obtained the functional law of the iterated logarithm in the sense of Strassen (1964) under condition (DMR) via the above Fuk–Nagaev type inequality and the coupling lemma of Chap. 5. Since the proof is rather technical, we will prove here only a bounded law of the iterated logarithm.

Throughout this section we use the notations $Lx = \log(x \vee e)$ and $LLx = L(Lx)$.

Theorem 6.4 *Let $(X_i)_{i>0}$ be a strictly stationary sequence of real-valued and centered random variables satisfying condition (DMR) for the sequence of strong mixing coefficients defined by (2.1). Then, with the same notation as in Theorem 6.2,*

$$\limsup_{n \to \infty} \frac{|S_n|}{s_n \sqrt{\log \log n}} \leq 8 \ \text{almost surely.}$$

Proof We first notice that, from the stationarity assumption,

$$\lim_{n\to\infty} n^{-1} s_n^2 = \operatorname{Var} X_0 + 2 \sum_{i=1}^{\infty} |\operatorname{Cov}(X_0, X_i)| = V > 0. \tag{6.23}$$

To prove Theorem 6.4, it is enough to prove that

$$\sum_{n>0} n^{-1} \mathbb{P}\left(\sup_{k\in[1,n]} |S_k| \geq 8 s_n \sqrt{LLn} \right) < \infty, \tag{6.24}$$

and next to apply the Borel–Cantelli lemma, as in Stout (1974, Chap. 5).

In order to prove (6.24), we now apply Theorem 6.2 with $r = 2LLn$ and $\lambda = \lambda_n = 2 s_n \sqrt{LLn}$. Let $x_n = \lambda/r = s_n (LLn)^{-1/2}$. Summing on n, we get that

$$\sum_{n>0} n^{-1} \mathbb{P}\left(\sup_{k\in[1,n]} |S_k| \geq 8 s_n \sqrt{LLn} \right) \leq 4 \sum_{n>0} \frac{1}{n} 3^{-LLn} + \sum_{n>0} \frac{4}{\lambda_n} \int_0^{H(x_n)} Q(u) du. \tag{6.25}$$

The series $\sum_{n>0} n^{-1} 3^{-LLn}$ is clearly convergent. To bound the second series on the right-hand side, we interchange the sum and integral: since $(u < H(x_n))$ if and only if $(R(u) > x_n)$, we thus obtain

$$\sum_{n>0} \frac{4}{\lambda_n} \int_0^1 Q(u) \mathbb{I}_{u<H(x_n)} du = 4 \int_0^1 Q(u) \left(\sum_{n>0} \frac{x_n}{s_n^2} \mathbb{I}_{x_n<R(u)} \right) du.$$

Now, by (6.23), the terms in the series are similar to $(nVLLn)^{-1/2}$. It follows that

$$\sum_{n>0} \frac{x_n}{s_n^2} \mathbb{I}_{x_n<R(u)} \leq C R(u)$$

for some positive constant C. Consequently,

$$\sum_{n>0} \frac{1}{\lambda_n} \int_0^{H(x_n)} Q(u) du \leq C \int_0^1 R(u) Q(u) du, \tag{6.26}$$

which implies (6.24). Hence Theorem 6.4 holds true. ∎

Exercises

(1) Let $(X_i)_{i>0}$ be a sequence of real-valued and centered random variables and $(\alpha_n)_{n\geq 0}$ be the sequence of strong mixing coefficients defined by (2.1). We assume that $s_n \geq 1$. Prove that if $\|X_i\|_\infty \leq 1$ for any positive i, then, for any λ in $[s_n, s_n^2]$,

$$\mathbb{P}\left(\sup_{k\in[1,n]}|S_k| \geq 4\lambda\right) \leq 4\exp\left(-\frac{\lambda^2}{4s_n^2}\right) + 4n\lambda^{-1}\alpha(s_n^2/\lambda). \tag{1}$$

Compare the terms on right-hand side of this inequality under the mixing assumption $\alpha_n = O(a^n)$ for some a in $]0, 1[$.

(2) *An inequality of Doukhan and Portal.* In this exercise, we will give an improved version of the exponential inequality of Doukhan and Portal (1987). We assume that $\|X_i\|_\infty \leq 1$ for any positive i and that the strong mixing coefficients defined by (2.1) satisfy $\alpha_q \leq c\exp(-aq)$ for any $q \geq 0$, for some positive constants a and c. Prove that, for any $n \geq 4$ and any $x \geq \log n$,

$$\mathbb{P}\left(|S_n| \geq 5(s_n \vee 2\sqrt{5})\sqrt{x} + \frac{10}{3a}x^2\right) \leq c\exp(-x). \tag{2}$$

(3) *Kolmogorov's law of the iterated logarithm.* Let $(X_i)_{i>0}$ be a sequence of identically distributed and independent centered random variables, with variance 1.
(a) Prove that, for any $\varepsilon > 0$ small enough,

$$\sum_{n>0} n^{-1}\mathbb{P}(S_n^* \geq (1+\varepsilon)^2\sqrt{2nLLn}) < \infty. \tag{3}$$

Hint: apply Theorem B.3(b) in Appendix B with $\lambda x = \varepsilon n$.
(b) Infer from (a) that

$$\limsup_{n\to\infty}(2nLLn)^{-1/2}|S_n| \leq 1 \text{ almost surely.}$$

Chapter 7
Empirical Distribution Functions

7.1 Introduction

In this chapter we are interested in functional limit theorems for the empirical distribution function associated to a stationary and strongly mixing sequence of random variables with values in $\mathbb{R}^d$. In the iid case, the functional central limit theorem for a suitably normalized and centered empirical distribution function is due to Donsker (1952). Donsker proved in particular that the Lipschitzian functionals of a suitably normalized and centered version of the empirical distribution function converge in distribution to the distribution of the corresponding functionals associated to a Brownian bridge. For this reason, the normalized and centered version of the empirical distribution function is often called the empirical bridge. Dudley (1966) extended the results of Donsker to the multivariate case, with a more rigorous approach. Following the approach of Dudley (1966), the proofs of these theorems generally include two steps. The first step consists in proving the finite-dimensional convergence of the empirical bridge to a suitable Gaussian process. The second step consists in proving the asymptotic equicontinuity of the empirical bridge for the uniform metric.

We now give a brief review of existing results before the year 2000 in the strong mixing case. Yoshihara (1979) extended the uniform central limit theorem of Donsker (1952) for the empirical bridge to stationary and strongly mixing sequences of real-valued random variables satisfying the strong mixing condition $\alpha_n = O(n^{-a})$ for some $a > 3$. Next Dhompongsa (1984) generalized Yoshihara's result to the multivariate case: he proved that, for random variables in $\mathbb{R}^d$, the uniform central limit theorem of Dudley (1966) for the multivariate distribution function holds if $\alpha_n = O(n^{-a})$ for some $a > d + 2$. Next Shao and Yu (1996) weakened the condition of Yoshihara (1979): they obtained Donsker's uniform central limit theorem under the strong mixing condition $\alpha_n = O(n^{-a})$ for some $a > 1 + \sqrt{2}$. In the β-mixing case, Arcones and Yu (1994), Doukhan et al. (1995) Rio (1998) obtained the uniform central limit theorem under slower rates of mixing in a more general setting. In particular, Rio (1998) proved that, for any $d \geq 1$, the uniform central limit theorem

© Springer-Verlag GmbH Germany 2017
E. Rio, *Asymptotic Theory of Weakly Dependent Random Processes*,
Probability Theory and Stochastic Modelling 80,
DOI 10.1007/978-3-662-54323-8_7

of Dudley (1966) for the multivariate distribution function holds if $\sum_{n>0} \beta_n < \infty$. Since this condition is the minimal β-mixing condition implying finite-dimensional convergence, this result cannot be improved. In Sect. 8.3 of Chap. 8, we will give another proof of this result. Nevertheless, the proofs in the β-mixing case involve coupling arguments and cannot be extended to the strong mixing case. In this section, we give less technical results in the strong mixing case, for the empirical distribution function. In particular, we will prove in Sects. 7.4 and 7.5 that the uniform central limit theorem of Dudley (1966) for the multivariate distribution function holds if $\alpha_n = O(n^{-a})$ for some $a > 1$. Before proving these theorems, we give in Sect. 7.2 an elementary L^2-estimate for the maximum of the empirical bridge. Next, in Sect. 7.3, we recall some facts from the theory of functional limit theorems. For further work on empirical distribution functions and empirical processes for dependent data, we refer to Dehling et al. (2002).

7.2   An Elementary Estimate

Let $(X_i)_{i \in \mathbb{Z}}$ be a sequence of real-valued random variables with common distribution function F. We set

$$F_n(x) = \frac{1}{n} \sum_{i=1}^{n} \mathbb{I}_{X_i \leq x} \text{ and } \nu_n(x) = \sqrt{n}(F_n(x) - F(x)). \tag{7.1}$$

The centered empirical measures P_n and Z_n are defined by (1.37). In this section we will study the rate of uniform convergence of F_n to F. Proposition 7.1 below provides an estimate of the L^2-norm of the maximal deviation. If the series of strong mixing coefficients is convergent, this estimate is optimal up to a logarithmic factor.

Proposition 7.1 *Let $(X_i)_{i \in \mathbb{Z}}$ be a strictly stationary sequence of real-valued random variables and let $(\alpha_k)_{k \geq 0}$ denote the sequence of strong mixing coefficients defined by (1.20). Suppose that the common distribution function F of the random variables is continuous. Then*

$$\mathbb{E}(\sup_{x \in \mathbb{R}} |\nu_n(x)|^2) \leq \left(1 + 4 \sum_{k=0}^{n-1} \alpha_k\right)\left(3 + \frac{\log n}{2 \log 2}\right)^2. \tag{7.2}$$

Proof For any Borel set A, let $I_n(A)$ be defined as in Exercise 5, Chap. 1. Let $(\varepsilon_i)_{i>0}$ be a sequence of independent and symmetric random variables with values in $\{-1, 1\}$. Then, for any finite partition $A_1, \ldots, A_k$ of A,

$$\sum_{i=1}^{k} \text{Var} Z_n(A_i) = \mathbb{E}\left(Z_n^2\left(\sum_{i=1}^{k} \varepsilon_i \mathbb{I}_{A_i}\right)\right), \tag{7.3}$$

which ensures that

$$I_n(A) \leq \sup\{\operatorname{Var} Z_n(f \, \mathbb{I}_A) : \|f\|_\infty \leq 1\}. \tag{7.4}$$

In order to prove Proposition 7.1, we now introduce a chaining argument. Let N be some positive integer, to be chosen later. For any real x such that $F(x) \neq 0$ and $F(x) \neq 1$, let us write $F(x)$ in base 2:

$$F(x) = \sum_{l=1}^{N} b_l(x) 2^{-l} + r_N(x) \text{ with } r_N(x) \in [0, 2^{-N}[,$$

where $b_l = 0$ or $b_L = 1$. For any L in $[1, N]$, set

$$\Pi_L(x) = \sum_{l=1}^{L} b_l(x) 2^{-l} \text{ and } i_L = \Pi_L(x) 2^L.$$

Let the reals $(x_L)_L$ be chosen in such a way that $F(x_L) = \Pi_L(x)$. With these notations

$$\nu_n(x) = \nu_n(\Pi_1(x)) + \sum_{L=2}^{N} \left(\nu_n(\Pi_L(x)) - \nu_n(\Pi_{L-1}(x)) \right) + \nu_n(x) - \nu_n(\Pi_N(x)). \tag{7.5}$$

Let the reals $x_{L,i}$ be defined by $F(x_{L,i}) = i2^{-L}$. From (7.5) we get that

$$\sup_{x \in [0,1]} |\nu_n(x)| \leq \sum_{L=1}^{N} \Delta_L + \Delta_N^*, \tag{7.6a}$$

with

$$\Delta_L = \sup_{i \in [1, 2^L]} |Z_n(]x_{L,i-1}, x_{L,i})| \text{ and } \Delta_N^* = \sup_{x \in \mathbb{R}} |Z_n(]\Pi_N(x), x])|. \tag{7.6b}$$

Let us now bound the L^2-norm of the maximum of the empirical process. By the triangle inequality,

$$\left(\mathbb{E}(\sup_{x \in [0,1]} |\nu_n(x)|^2) \right)^{1/2} \leq \sum_{L=1}^{N} \|\Delta_L\|_2 + \|\Delta_N^*\|_2. \tag{7.7}$$

Since

$$\Delta_L^2 \leq \sum_{i=1}^{2^L} Z_n^2(](i-1)2^{-L}, i2^{-L}]),$$

it follows from both (7.4) and Theorem 1.1 that

$$\mathbb{E}(\Delta_L^2) \leq \sum_{i=1}^{2^L} \operatorname{Var} Z_n(](i-1)2^{-L}, i2^{-L}]) \leq 1 + 4 \sum_{k=0}^{n-1} \alpha_k. \qquad (7.8)$$

It remains to bound Δ_N^*. From the inequalities

$$-\sqrt{n}2^{-N} \leq Z_n(]\Pi_N(x), x]) \leq Z_n(]\Pi_N(x), \Pi_N(x) + 2^{-N}]) + \sqrt{n}2^{-N},$$

we get that

$$\Delta_N^* \leq \Delta_N + \sqrt{n}2^{-N}. \qquad (7.9)$$

The inequalities (7.7)–(7.9) then ensure that

$$\left(\mathbb{E}(\sup_{x \in [0,1]} |\nu_n(x)|^2)\right)^{1/2} \leq (1 + N + \sqrt{n}\,2^{-N})\left(1 + 4 \sum_{k=0}^{n-1} \alpha_k\right)^{1/2}. \qquad (7.10)$$

Taking $N = 1 + [(2 \log 2)^{-1} \log n]$ and noticing that $\sqrt{n}2^{-N} \leq 1$ for this choice of N, we then get Proposition 7.1. ∎

7.3   Functional Central Limit Theorems

In Sect. 7.2, we proved that, under the strong mixing condition (1.24), the order of magnitude of the supremum of the empirical bridge is at most $O(\log n)$. Now, if the strong mixing coefficients are defined by (2.1), the mixing condition $\sum_n \alpha_n < \infty$ implies the finite-dimensional convergence of the empirical bridges ν_n to a Gaussian process G with covariance function

$$\operatorname{Cov}(G(x), G(y)) = \sum_{t \in \mathbb{Z}} \operatorname{Cov}(\mathbb{I}_{X_0 \leq x}, \mathbb{I}_{X_t \leq y}). \qquad (7.11)$$

Here we are interested in the uniform convergence with respect to x of ν_n to G. Such a result will be called a uniform central limit theorem or functional central limit theorem. In this section we give a precise definition of the notion of a uniform central limit theorem and sufficient conditions for the theorem to hold. Our exposition is derived from Pollard (1990, Sect. 10).

Let (T, ρ) be a metric or a pseudo-metric space. Denote by $B(T)$ the space of real-valued and bounded functions on T. On $B(T)$ we consider the uniform distance

$$d(x, y) = \sup_{t \in T} |x(t) - y(t)|.$$

Let $\{X_n(\omega, t) : t \in T\}$ be a sequence of real-valued random processes on T. We are interested in the convergence in distribution of this sequence under the distance d. More precisely, we have in view the functional convergence to a Gaussian process with trajectories in the space

$$U_\rho(T) = \{x \in B(T) : x \text{ is uniformly continuous under } \rho\}.$$

Under the distance d, the space $U_\rho(T)$ is countably generated if and only if (T, ρ) is totally bounded. In that case, a Borel probability measure P on $U_\rho(T)$ is uniquely determined by the finite-dimensional projections

$$P(B \mid t_1, \ldots, t_k) = P\{x \in U_\rho(T) : (x(t_1), \ldots, x(t_k)) \in B\},$$

where $\{t_1, \ldots t_k\}$ ranges over the set of finite subsets of T and B is any Borel subset of $\mathbb{R}^k$.

For example, in the case of random variables with uniform distribution over $[0, 1]$, the space $T = [0, 1]$ is equipped with the usual distance on $\mathbb{R}$. Then the Gaussian process G with covariance function defined by (7.11) is uniquely defined as soon as its law is concentrated on $U_\rho(T)$.

We now recall the definition of finite-dimensional convergence (fidi convergence). The fidi convergence of $(X_n(\cdot, t))$ holds true if and only if for any finite subset $\{t_1, \ldots t_k\}$ of T there exists a probability measure P such that

$$(X_n(\cdot, t_1), \ldots, (X_n(\cdot, t_k)) \longrightarrow P(\cdot \mid t_1, \ldots, t_k) \text{ in distribution.} \qquad (7.12)$$

We now give a criterion for the convergence in $U_\rho(T)$

Theorem 7.1 *Theorem 10.2 in Pollard (1990) - Let (T, ρ) be a totally bounded pseudo-metric space and let $\{X_n(\omega, t) : t \in T\}$ be a sequence of random processes on T. Suppose that*

(i) The fidi convergence in the sense of (7.12) holds true.
(ii) For any positive ε and η, there exists a positive δ such that

$$\limsup_{n \to \infty} \mathbb{P}^* \left\{ \sup_{\substack{(s,t) \in T \times T \\ \rho(s,t) < \delta}} |X_n(\omega, s) - X_n(\omega, t)| > \eta \right\} < \varepsilon.$$

Then there exists a Borel probability measure P concentrated on $U_\rho(T)$ with finite-dimensional margins given by (7.12). Furthermore, X_n converges in distribution to P in the space $B(T)$.

Conversely, if X_n converges in distribution to a probability measure P concentrated on $U_\rho(T)$, then conditions (i) and (ii) are fulfilled.

Condition (ii) is called stochastic equicontinuity. If the limiting process is a Gaussian process then (X_n) is said to satisfy the functional central limit theorem or the uniform central limit theorem. We refer to Pollard (1990) for a proof of this result. Now, in Sect. 7.4 below we apply this result to the functional central limit theorem for the empirical bridge in the strong mixing case.

7.4   A Functional Central Limit Theorem for the Empirical Distribution Function

In this section, we prove a functional central limit theorem for the empirical distribution function associated to a stationary strongly mixing sequence of real-valued random variables. In order to give elementary proofs, we will assume that the common distribution function of the random variables is continuous. Nevertheless, this result can be extended to arbitrary distribution functions. Theorem 7.2 below improves previous results of Yoshihara (1979) and Shao and Yu (1996). We refer to Doukhan and Surgailis (1998), Louhichi (2000), Dehling et al. (2009) and (Dedecker, 2010) for other types of dependence.

Theorem 7.2 *Let $(X_i)_{i \in \mathbb{Z}}$ be a strictly stationary sequence of real-valued random variables with common continuous distribution function F. Assume that the sequence $(\alpha_n)_{n \geq 0}$ of strong mixing coefficients defined by (2.1) satisfies*

$$\alpha_n \leq cn^{-a} \text{ for some real } a > 1 \text{ and some constant } c \geq 1. \qquad (i)$$

Then there exists a Gaussian process G with uniformly continuous trajectories on $\mathbb{R}$ equipped with the pseudo-metric d_F given by $d_F(x, y) = |F(x) - F(y)|$, such that ν_n converges in distribution to G in $B(\mathbb{R})$ as n tends to ∞.

Proof Considering $U_i = F(X_i)$ it is sufficient to prove Theorem 7.2 for random variables with the uniform distribution over $[0, 1]$. Now, by Corollary 4.1, the fidi convergence to a Gaussian process G with covariance defined by (7.11) holds. According to Theorem 7.2 it remains to prove the stochastic equicontinuity property (ii). This property follows immediately from the proposition below.

Proposition 7.2 *Let $(X_i)_{i \in \mathbb{Z}}$ be a strictly stationary sequence of random variables with uniform distribution over $[0, 1]$. Assume that $(X_i)_{i \in \mathbb{Z}}$ satisfies the strong mixing condition (i) of Theorem 7.2. Let $\Pi_K(x) = 2^{-K}[2^K x]$. Then*

$$\lim_{K \to \infty} \limsup_{n \to \infty} \mathbb{E}^* \left(\sup_{x \in [0,1]} |\nu_n(x) - \nu_n(\Pi_K(x))| \right) = 0.$$

Proof of Proposition 7.2. Proceeding as in the proof of (7.6), we first obtain that

$$\sup_{x \in [0,1]} |\nu_n(x) - \nu_n(\Pi_K(x))| \leq \sum_{L=K+1}^{N} \Delta_L + \Delta_N^*.$$

Now, applying (7.9), we have:

$$\sup_{x \in [0,1]} |\nu_n(x) - \nu_n(\Pi_K(x))| \leq \sum_{L=K+1}^{N} \Delta_L + \Delta_N + \sqrt{n}\, 2^{-N} := \Delta. \qquad (7.13)$$

By the triangle inequality,

$$\|\Delta\|_1 \leq \sqrt{n}\, 2^{-N} + \sum_{L=K+1}^{N-1} \|\Delta_L\|_1 + 2\|\Delta_N\|_1. \qquad (7.14)$$

Let N be the natural number such that $2^{N-1} < n \leq 2^N$. For this choice of N, by (7.14),

$$\|\Delta\|_1 \leq n^{-1/2} + 2 \sum_{L=K+1}^{N} \|\Delta_L\|_1. \qquad (7.15)$$

Hence Proposition 7.2 follows from the lemma below.

Lemma 7.1 *Let N be the natural number such that $2^{N-1} < n \leq 2^N$. Then there exists a positive constant C_0 depending only on a and c such that*

$$\|\Delta_L\|_1 \leq C_0 2^{-(a-1)^2 L/(4a)^2} \text{ for any } L \in [1, N].$$

Proof of Lemma 7.1. Define the dyadic intervals $I_{L,i}$ by $I_{L,i} =](i-1)2^{-L}, i2^{-L}]$ for any integer i in $[1, 2^L]$. In order to prove Lemma 7.1, we will refine the symmetrization technique introduced in Sect. 7.2. As in Sect. 7.2, let $(\varepsilon_i)_{i \in [1,2^L]}$ be a sequence of independent symmetric signs, independent of the sequence $(X_i)_{i \in \mathbb{Z}}$.

Let J be a finite subset of integers in $[1, 2^L]$. Assume that the supremum of the random variables $|Z_n(I_{L,i})|$ as i ranges over J is more than x. Let j be the smallest integer in J such that $|Z_n(I_{L,i})| \geq x$. Then, for any choice of the signs $(\varepsilon_i)_{i \in J \setminus \{j\}}$, either

$$Z_n\left(\sum_{i \in J \setminus \{j\}} \varepsilon_i\, \mathbb{1}_{I_{L,i}}\right) + Z_n(I_{l,j}) \text{ or } Z_n\left(\sum_{i \in J \setminus \{j\}} \varepsilon_i\, \mathbb{1}_{I_{L,i}}\right) - Z_n(I_{l,j})$$

does not belong to the interval $]-x, x[$. Consequently,

$$\text{Card}\left\{i \in J \text{ such that } \left|Z_n\left(\sum_{i \in J} \varepsilon_i\, \mathbb{1}_{I_{l,i}}\right)\right| \geq x\right\} \geq 2^{|J|-1},$$

whence

$$\mathbb{P}\left(\sup_{i \in J} |Z_n(I_{L,i})| \geq x\right) \leq 2\mathbb{P}\left(\left|Z_n\left(\sum_{i \in J} \varepsilon_i\, \mathbb{1}_{I_{L,i}}\right)\right| \geq x\right). \qquad (7.16)$$

Let M be an integer in $[1, L]$, to be chosen later. For any k in $[1, 2^M]$, let

$$J_k = \{(k-1)2^{L-M} + 1, \ldots, k2^{L-M}\}.$$

Applying (7.16), we obtain that

$$\mathbb{P}(\Delta_L \geq x) \leq 2 \sum_{k=1}^{2^M} \mathbb{P}\left(\left|Z_n\left(\sum_{i \in J_k} \varepsilon_i \,\mathbb{1}_{I_{L,i}}\right)\right| \geq x\right). \tag{7.17}$$

Throughout the sequel, C denotes a positive constant depending on a and c, which may change from line to line. Let us fix the values of the signs ε_i. Applying Corollary 1.1 to the random variables $Y_l = \sum_{i \in J_k} \varepsilon_i \,\mathbb{1}_{I_{L,i}}(X_l)$, we have:

$$\sum_{l=1}^{n} \sum_{m=1}^{n} |\mathrm{Cov}(Y_l, Y_m)| \leq 4 \int_0^{2^{-M}} \alpha^{-1}(u)du \leq 4c \sum_{i=0}^{\infty} \inf(i^{-a}, 2^{-M}) \leq C2^{M(1-a)/a}. \tag{7.18}$$

Therefrom, applying inequality (6.19b) to the random variables $Z_n\left(\sum_{i \in J_k} \varepsilon_i \,\mathbb{1}_{I_{L,i}}\right)$ conditionally to the values of the signs,

$$\mathbb{P}\left(\left|Z_n\left(\sum_{i \in J_k} \varepsilon_i \,\mathbb{1}_{I_{L,i}}\right)\right| \geq 4\lambda\right) \leq Cr^{r/2} 2^{M(1-a)r/(2a)}\lambda^{-r} + 2c(2r)^{a+1}n^{(1-a)/2}\lambda^{-a-1}.$$

Now, by (7.17),

$$\begin{aligned}
\mathbb{P}(\Delta_L \geq 4\lambda) \leq \; &Cr^{r/2} \min(1, 2^{M(2a+(1-a)r)/(2a)}\lambda^{-r}) \\
&+ 2c(2r)^{a+1} \min(1, 2^M n^{(1-a)/2}\lambda^{-a-1}).
\end{aligned} \tag{7.19}$$

Let $r = 4a/(a-1)$. For this value of r, inequality (7.19) yields

$$\mathbb{P}(\Delta_L \geq 4\lambda) \leq C \min(1, 2^{-M}\lambda^{-r}) + C \min(1, 2^M n^{(1-a)/2}\lambda^{-a-1}). \tag{7.20}$$

Integrating (7.20) with respect to λ, we get that

$$\mathbb{E}(\Delta_L) \leq 8C\left(2^{-M/r} + 2^{M/(a+1)}n^{(1-a)/(2a+2)}\right). \tag{7.21}$$

Now, choosing $M = [L(a-1)/(4a)] = [L/r]$ and noticing that $n \geq 2^{L-1}$, we infer from (7.21) that

$$\mathbb{E}(\Delta_L) \leq 16C\left(2^{-L/r^2} + 2^{-L(2a-1)/(ra+r)}\right) \leq 32C2^{-L/r^2}, \tag{7.22}$$

which implies Lemma 7.1. Hence Proposition 7.2 holds, which completes the proof of Theorem 7.2. ∎

7.5 Multivariate Distribution Functions

Throughout this section, $\mathbb{R}^d$ is equipped with the usual product order. Let $(X_i)_{i \in \mathbb{Z}}$ be a strictly stationary sequence of random variables with values in $\mathbb{R}^d$. We set

$$F_n(x) = n^{-1} \sum_{i=1}^{n} \mathbb{1}_{X_i \leq x} \text{ and } F(x) = \mathbb{P}(X_0 \leq x).$$

The so-defined empirical distribution function corresponds to the empirical process indexed by the class of lower-left closed orthants. We then define the empirical bridges ν_n by

$$\nu_n(x) = \sqrt{n} \left(F_n(x) - F(x) \right).$$

The result below extends Theorem 7.2 to the multivariate case. The most striking fact is that, for multivariate distribution functions, the mixing condition does not depend on the dimension d, in contrast to the previous results on the same subject. We refer to Bücher (2015) for an extension of this result to the sequential empirical process.

Theorem 7.3 *Let $(X_i)_{i \in \mathbb{Z}}$ be a strictly stationary sequence of random variables with values in $\mathbb{R}^d$. For each j in $[1, d]$, let F_j denote the distribution function of the j-th component of X_0. Suppose that the distribution functions F_j are continuous. Assume furthermore that the strong mixing condition (i) of Theorem 7.2 holds true for the strong mixing coefficients defined by (2.1). Then there exists a Gaussian process G with uniformly continuous trajectories on $\mathbb{R}^d$ equipped with the pseudo-metric d_F given by*

$$d_F(x, y) = \sup_{j \in [1, d]} |F_j(x_j) - F_j(y_j)|, \text{ where } x = (x_1, \ldots x_d) \text{ and } y = (y_1, \ldots y_d),$$

such that ν_n converges in distribution to G in $B(\mathbb{R}^d)$ as n tends to ∞.

Proof Let $X_i = (X_i^1, \ldots, X_i^d)$. Define the random variables Y_i in $[0, 1]^d$ by $Y_i = (F_1(X_i^1), \ldots, F_d(X_i^d))$. Since the marginal distribution functions F_i are continuous, the random variables Y_i have uniform margins. Consequently, in order to prove Theorem 7.3, we may assume, without loss of generality, that the random variables X_i are with values in $[0, 1]^d$ and with marginal distributions the uniform distribution over $[0, 1]$. In that case d_F is the distance induced by the norm $\| . \|_\infty$. Under condition (i) of Theorem 7.2, the strong mixing coefficients are summable. Hence Corollary 4.1 implies the fidi convergence of a Gaussian process with covariance defined by (7.11). In view of Theorem 7.1, it remains to prove the stochastic equicontinuity property. This property follows immediately from Proposition 7.3 below.

Proposition 7.3 *Let $(X_i)_{i \in \mathbb{Z}}$ be a strictly stationary sequence of random variables with values in $[0, 1]^d$. Suppose that the coordinates of X_0 have the uniform distribution over $[0, 1]$. For any $x = (x_1, \ldots x_d)$ in $\mathbb{R}^d$ and any positive integer K,* *let* $\Pi_K(x) = (2^{-K}[2^K x_1], \ldots, 2^{-K}[2^K x_d])$. *Then, under the assumptions of Theorem 7.3,*

$$\lim_{K \to +\infty} \limsup_{n \to \infty} \mathbb{E}^* \left(\sup_{x \in [0,1]^d} |\nu_n(x) - \nu_n(\Pi_K(x))| \right) = 0.$$

Proof Let N be the unique integer such that $2^{N-1} < n \le 2^N$. Clearly

$$\nu_n(x) = \nu_n(x) - \nu_n(\Pi_N(x)) + \nu_n(\Pi_N(x)).$$

Hence,

$$\sup_{x \in [0,1]^d} |\nu_n(x) - \nu_n(\Pi_K(x))| \le \sup_{x \in [0,1]^d} |\nu_n(\Pi_N(x)) - \nu_n(\Pi_K(x))| + R_N \quad (7.23)$$

with

$$R_N = \sup_{x \in [0,1]^d} |\nu_n(x) - \nu_n(\Pi_N(x))|.$$

In order to bound R_N, we will use the elementary result below.

Lemma 7.2 *Let μ be a probability measure on $\mathbb{R}^d$ with distribution function G. For each j in $[1, d]$, let G_j denote the distribution function of the j-th marginal of μ, which is defined by $G_j(x) = \mu(\mathbb{R}^{j-1} \times] - \infty, x] \times \mathbb{R}^{d-j})$. Then, for any $x = (x_1, \ldots x_d)$ and any $y = (y_1, \ldots y_d)$ in $\mathbb{R}^d$,*

$$|G(x) - G(y)| \le \sum_{j=1}^d \left(G_j(x_j \vee y_j) - G_j(x_j \wedge y_j) \right).$$

Proof of Lemma 7.2. Let $\mathcal{Q}_x = \{z \in \mathbb{R}^d \text{ such that } z \le x\}$. If Δ denotes the symmetric difference, then

$$|G(x) - G(y)| = |\mu(\mathcal{Q}_x) - \mu(\mathcal{Q}_y)| \le \mu(\mathcal{Q}_x \Delta \mathcal{Q}_y).$$

Now $\mathcal{Q}_x \Delta \mathcal{Q}_y \subset \bigcup_{j=1}^d \mathbb{R}^{j-1} \times]x_j \wedge y_j, x_j \vee y_j] \times \mathbb{R}^{d-j}$, which, together with the subadditivity of μ and the above inequality implies Lemma 7.2. ∎

Using Lemma 7.2 we now bound R_N. Let $F_{n,j}$ denote the empirical distribution function associated to the j-th components X_i^j of the random variables X_i, which is defined by $F_{n,j}(x_j) = F_n(1, \ldots, 1, x_j, 1, \ldots, 1)$ and let $\nu_{n,j}(x_j) = \sqrt{n}(F_{n,j}(x_j) - F_j(x_j))$. By Lemma 7.2 applied twice,

$$R_N \le \sqrt{n} \sup_{x \in [0,1]^d} \left(\sum_{j=1}^d (F_{n,j}(x_j) - F_{n,j}(\Pi_N(x_j)) + x_j - \Pi_N(x_j)) \right).$$

Now $x_j - \Pi_N(x_j) \le 2^{-N}$, which ensures that

$$R_N \le \sqrt{n}\, d 2^{-N} + \sqrt{n} \sup_{x\in[0,1]^d} \sum_{j=1}^{d} \big(F_{n,j}(x_j) - F_{n,j}(\Pi_N(x_j))\big).$$

Next, from the monotonicity properties of the empirical distribution functions $F_{n,j}$,

$$\sqrt{n}\Big(F_{n,j}(x_j) - F_{n,j}(\Pi_N(x_j))\Big) \le \sqrt{n}\Big(F_{n,j}(\Pi_N(x_j) + 2^{-N}) - F_{n,j}(\Pi_N(x_j))\Big)$$
$$\le \sqrt{n}\, 2^{-N} + \nu_{n,j}(\Pi_N(x_j) + 2^{-N}) - \nu_{n,j}(\Pi_N(x_j)).$$

Since $2^N \ge n$, it follows that

$$R_N \le 2dn^{-1/2} + \sum_{j=1}^{d} \sup_{x_j\in[0,1]} \Big(\nu_{n,j}(\Pi_N(x_j) + 2^{-N}) - \nu_{n,j}(\Pi_N(x_j))\Big).$$

Now the sequence of real-valued random variables $(X_i^j)_{i>0}$ still satisfy the strong mixing condition (i). Hence Lemma 7.1 can be applied with $L = N$ to each of the random variables in the sum on right-hand side, yielding

$$\mathbb{E}(R_N) \le 2dn^{-1/2} + dC_0 n^{-(a-1)^2/(4a)^2}. \tag{7.24}$$

We now bound the main term in (7.23). For any $x = (x_1, \dots, x_d)$ in the unit cube $]0, 1]^d$, let $]0, x] =]0, x_1] \times \cdots \times]0, x_d]$. For any j in $[1, d]$ and any natural integer M,

$$]0, \Pi_M(x_j)] = \bigcup_{L_j=0}^{M}]\Pi_{L_j-1}(x_j), \Pi_{L_j}(x_j)]$$

(note that $\Pi_{-1}(x_j) = 0$). Hence, taking the product,

$$]0, \Pi_M(x)] = \bigcup_{L\in[0,M]^d} \prod_{i=j}^{d}]\Pi_{L_j-1}(x_j), \Pi_{L_j}(x_j)].$$

Consequently,

$$]0, \Pi_N(x)]\backslash]0, \Pi_K(x)] = \bigcup_{\substack{L\in[0,N]^d \\ L\notin[0,K]^d}} \prod_{j=1}^{d}]\Pi_{L_j-1}(x_j), \Pi_{L_j}(x_j)]. \tag{7.25}$$

Notation 7.1 For any $L = (L_1, \dots, L_d)$ in $\mathbb{N}^d$, let $\mathcal{D}_L$ be the class of dyadic boxes $\prod_{i=1}^{d}](k_i-1)2^{-L_i}, k_i 2^{-L_i}]$ (here $k = (k_1, \dots, k_d)$ are multivariate natural numbers).

Let $Z_n = \sqrt{n}(P_n - P)$ denote the normalized and centered empirical measure, as defined in (1.34). We set

$$\Delta_L = \sup_{S \in \mathcal{D}_L} |Z_n(S)|.$$

With these notations,

$$\Delta := \sup_{x \in [0,1]^d} |\nu_n(\Pi_N(x)) - \nu_n(\Pi_K(x))| \leq \sum_{L \in [0,N]^d \setminus [0,K]^d} \Delta_L. \qquad (7.26)$$

For a fixed L we now consider the smallest integer j such that $L_j = \max(L_1, \ldots, L_d)$. Suppose, for example, that $j = 1$. Let M be a fixed integer in $[1, L_1]$, and k in $[1, 2^M]$, and let J_k be the set of elements of $\mathcal{D}_L$ contained in the strip $(k-1)2^{-M} < x_1 \leq k2^{-M}$.

We now adapt the symmetrization method of Sect. 7.4 to the multivariate case. Let $(\varepsilon_S)_{S \in \mathcal{D}_L}$ be a sequence of independent symmetric signs, independent of the sequence $(X_i)_{i \in \mathbb{Z}}$. Inequality (7.17) still holds in the multivariate case, and has the following structure:

$$\mathbb{P}(\Delta_L \geq x) \leq 2 \sum_{k=1}^{2^M} \mathbb{P}\left(\left|Z_n\left(\sum_{S \in J_k} \varepsilon_S \mathbb{1}_S\right)\right| \geq x\right).$$

Now

$$\left|\sum_{S \in J_k} \varepsilon_S \mathbb{1}_S(X_i)\right| = \mathbb{1}_{X_i^1 \in I_{M,k}},$$

and consequently (7.18) remains true (recall that the random variables X_i^1 are uniformly distributed over $[0, 1]$). Next, applying Inequality (6.19b) with $r = 4a/(a-1)$ as in Sect. 7.4, we get that, for any M in $[1, \max(L_1, \ldots, L_d)]$,

$$\mathbb{P}(\Delta_L \geq 4\lambda) \leq C \min(1, 2^{-M}\lambda^{-r}) + C \min(1, 2^M n^{(1-a)/2}\lambda^{-a-1}). \qquad (7.27)$$

Let $|L|_\infty = \max(L_1, \ldots, L_d)$ and choose $M = \lceil |L|_\infty / r \rceil$. Since $n \geq 2^{N-1} \geq 2^{|L|_\infty - 1}$, integrating (7.27) with respect to λ, we get that

$$\mathbb{E}(|\Delta_L|) \leq 32C2^{-\theta |L|_\infty}, \quad \text{with } \theta = r^{-2} = (a-1)^2/(16a^2). \qquad (7.28)$$

Now the cardinality of the set of integers L in $\mathbb{N}^d$ such that $|L|_\infty = J$ is exactly $(J+1)^d - J^d$. Consequently, the inequalities (7.23), (7.24), (7.26) and (7.28) yield

$$\mathbb{E}^*\left(\sup_{x \in [0,1]^d} |\nu_n(x) - \nu_n(\Pi_K(x))|\right) \leq d\left(32C \sum_{J>K} (J+1)^d 2^{-J\theta} + C_0 dn^{-\theta} + 2dn^{-1/2}\right), \qquad (7.29)$$

which implies Proposition 7.3 and, consequently, Theorem 7.3. ∎

Chapter 8
Empirical Processes Indexed by Classes of Functions

8.1 Introduction

In this chapter, we give new uniform central limit theorems for general empirical processes indexed by classes of sets or classes of functions. In Sect. 8.2, we consider convex sets of functions embedded in spaces of regular functions. In that case, the conditions implying the uniform central limit theorem are described in terms of regularity of the functions. Here the theory of approximation of functions (see Devore and Lorentz (1993)) is a fundamental tool. This tool is used to obtain the stochastic equicontinuity in Theorem 8.1 under the minimal strong mixing condition $\sum_k \alpha_k < \infty$. This result is similar to previous results of Doukhan et al. (1984) and Massart (1987) for classes of regular functions.

In Sect. 8.3, we give new results for empirical processes indexed by absolutely regular sequences. Arcones and Yu (1994) and Doukhan et al. (1995) give extensions of the results of Pollard (1982) and Ossiander (1987) to absolutely regular sequences. Nevertheless, these results still lead to suboptimal applications: for example, the uniform central limit theorem holds for the normalized and centered multivariate empirical distribution function as soon as the β-mixing coefficients satisfy $\beta_n = O(n^{-b})$ for some $b > 1$. By contrast, Rio (1998) obtains the uniform central limit theorem for the multivariate empirical distribution function and more generally for empirical processes indexed by Vapnik–Chervonenkis classes of sets under the minimal absolute regularity condition $\sum_{i>0} \beta_i < \infty$. The proof of Rio (1998) is based on the maximal coupling theorem of Goldstein (1979). In Sect. 8.3, we will adapt the proof of Rio (1998) to classes of functions satisfying bracketing conditions. Again the results of Sect. 8.3 yield the uniform central limit theorem for the multivariate empirical distribution function under the minimal regularity condition $\sum_{i>0} \beta_i < \infty$, in contrast to the results of Arcones and Yu (1994) and Doukhan et al. (1995).

© Springer-Verlag GmbH Germany 2017
E. Rio, *Asymptotic Theory of Weakly Dependent Random Processes*,
Probability Theory and Stochastic Modelling 80,
DOI 10.1007/978-3-662-54323-8_8

8.2   Classes of Regular Functions

In this section, we are interested in convex subsets of classes of regular functions. We
will prove in Proposition 8.1 that, for unit balls of some spaces of regular functions,
the stochastic equicontinuity property holds true for the empirical process. We then
derive from this first result a uniform central limit theorem for the empirical process
indexed by compact subsets of this unit ball.

Here we will consider generalized Lipschitz spaces, such as the Zygmund space.
We start with some definitions and elementary properties of these spaces. We refer
to the books of Meyer (1990) and Devore and Lorentz (1993) for more about these
spaces and their properties. In order to define these spaces, we need to introduce the
integrated modulus of regularity. For the sake of brevity, we give the definition only
in the real case.

Definition 8.1 For any real t, let T_h be the shift operator which maps the function
f to the function $T_h f$, which is defined by $T_h f(x) = f(x + h)$ for any x. Let

$$\Delta_h^r(f, x) = (T_h - T_0)^r f(x).$$

Let p be any real in $[1, +\infty]$ ($p = \infty$ is included). For any closed subinterval I of
$\mathbb{R}$ and any function f in $L^p(I)$, we define the integrated modulus of regularity of
order r of f by

$$\omega_r(f, t)_p = \sup_{h \in]0,t]} \left(\int_{I_{rh}} |\Delta_h^r(f, x)|^p dx \right)^{1/p},$$

where I_{rh} is the closed interval such that $\inf I_{rh} = \inf I$ and $\sup I_{rh} = \sup I - rh$.

We now define the generalized Lipschitz spaces of order s in the univariate case.

Definition 8.2 Let s be any positive real. Set $r = [s] + 1$. We denote by $\mathrm{Lip}^*(s, p, I)$
the space of functions f in $L^p(I)$ such that for some positive constant M,

$$\left(\int_{I_{rh}} |\Delta_h^r(f, x)|^p dx \right)^{1/p} \leq M h^s \quad \text{for any } h > 0.$$

On $\mathrm{Lip}^*(s, p, I)$, we consider the semi-norm

$$|f|_{\mathrm{Lip}^*(s,p)} = \sup_{t>0} t^{-s} \omega_r(f, t)_p.$$

We define a norm on $\mathrm{Lip}^*(s, p, I)$ by $\|f\|_{\mathrm{Lip}^*(s,p)} = |f|_{\mathrm{Lip}^*(s,p)} + \|f\|_p$. Let
$B(s, p, I)$ denote the unit ball associated to this norm.

Remark 8.1 In the case $p = 2$, the space $\mathrm{Lip}^*(s, 2, \mathbb{R})$ contains the Sobolev space
of order s. For $s = 1$ and $p = \infty$, the space $\mathrm{Lip}^*(1, \infty, I)$ is the Zygmund space
$Z(I)$ of functions f such that $|f(x + 2t) - 2f(x + t) + f(x)| \leq Mt$. This space
contains the space $\mathrm{Lip}(1, \infty, I)$ of Lipschitz functions on I.

In order to prove the stochastic equicontinuity property for empirical processes indexed by balls of these classes of functions, it will be convenient to use the wavelet expansions of these functions. Below we give the characterization of the spaces Lip*$(s, p, \mathbb{R}^d)$ for any $d \geq 1$. We refer to Meyer (1990, T. 1., pp. 196–198) for a definition of these spaces in the multivariate case and for more about wavelets.

For any nonnegative integer j, let us consider $\Lambda_j = 2^{-j-1}\mathbb{Z}^d \setminus 2^{-j}\mathbb{Z}^d$. Let us define $\Lambda = \mathbb{Z}^d \bigcup (\bigcup_{j\in\mathbb{N}} \Lambda_j)$. We consider a multiresolution analysis in $L^2(\mathbb{R}^d)$. For λ in $\Lambda \setminus \mathbb{Z}^d$, we denote by ψ_λ the wavelet of the multiresolution analysis corresponding to λ. Then the wavelets $\{\psi_\lambda : \lambda \in \Lambda_j\}$ form an orthonormal system. For $j > 0$, we denote by W_j the subspace of $L^2(\mathbb{R}^d)$ generated by this system. Let φ denote the father function. For λ in $\mathbb{Z}^d$, we set $\varphi_\lambda(x) = \varphi(x - \lambda)$ and we denote by V_0 the subspace of $L^2(\mathbb{R}^d)$ generated by the orthonormal system $\{\varphi_\lambda : \lambda \in \mathbb{Z}^d\}$. Then

$$L^2(\mathbb{R}^d) = V_0 \overset{\perp}{\oplus} \left(\overset{\perp}{\underset{j>0}{\oplus}} W_j \right).$$

Throughout the sequel we assume that the scaling functions have a compact support and are $1 + [s]$ times continuously differentiable. For the sake of convenience, we set $\psi_\lambda = \varphi_\lambda$ for λ in $\mathbb{Z}^d$. Then any function f in $L^2(\mathbb{R}^d)$ has the orthogonal expansion

$$f = \sum_{\lambda\in\mathbb{Z}^d} a_\lambda \varphi_\lambda + \sum_{j=0}^{\infty} \sum_{\lambda\in\Lambda_j} a_\lambda \psi_\lambda = \sum_{\lambda\in\Lambda} a_\lambda \psi_\lambda. \tag{8.1}$$

Let Lip*$(s, p, \mathbb{R}^d)$ denote the generalized Lipschitz spaces, as defined in Meyer (1990), and let $B(s, p, \mathbb{R}^d)$ denote the unit ball of this space. Meyer (1990) gives the following characterization of these spaces.

Proposition 8.1 *For f in* Lip*$(s, p, \mathbb{R}^d)$, *let*

$$\|f\|_{ond} = \sup\left(\left(\sum_{\lambda\in\mathbb{Z}^d} |a_\lambda|^p \right)^{1/p}, \sup_{j\in\mathbb{N}} \left(\left(\sum_{\lambda\in\Lambda_j} |a_\lambda|^p \right)^{1/p} 2^{js+jd/2-jd/p} \right) \right).$$

Then $\| . \|_{ond}$ is a norm on Lip*$(s, p, \mathbb{R}^d)$. *Furthermore, this norm is equivalent to the usual norm $\| . \|_{\text{Lip}^*(s,p,\mathbb{R}^d)}$ on* Lip*$(s, p, \mathbb{R}^d)$.

In order to compare these spaces, it will be convenient to use the elementary result below.

Lemma 8.1 *Let K be a countable set and $(a_k)_{k\in K}$ be a family of nonnegative reals. Then for any reals $q > p > 0$,*

$$\left(\sum_{k\in K} a_k^q \right)^{1/q} \leq \left(\sum_{k\in K} a_k^p \right)^{1/p}.$$

By Lemma 8.1, for any p in $[1, 2]$,

$$\left(\sum_{\lambda \in \Lambda_j} a_\lambda^2\right)^{1/2} \le \|f\|_{ond} 2^{j(-s-d/2+d/p)} \quad \text{and} \quad \left(\sum_{\lambda \in \mathbb{Z}^d} a_\lambda^2\right)^{1/2} \le \|f\|_{ond}. \tag{8.2}$$

It follows from (8.2) that $\text{Lip}^*(s, p, \mathbb{R}^d) \subset L^2(\mathbb{R}^d)$ for $s > (d/p) - (d/2)$. Moreover, $\text{Lip}^*(s, p, \mathbb{R}^d) \subset L^\infty(\mathbb{R}^d)$ for $s > d/p$. Let us now give the main result of this section.

Theorem 8.1 *Let $(X_i)_{i \in \mathbb{Z}}$ be a strictly stationary sequence of random variables with values in $\mathbb{R}^d$. Suppose that the strong mixing coefficients $(\alpha_k)_{k \ge 0}$ defined by (1.20) satisfy $\sum_{k \ge 0} \alpha_k < \infty$. Let p be any real in $[1, 2]$ and let s be any real such that $s > d/p$. Let a be a positive real and let $\mathcal{F} = a B(s, p, \mathbb{R}^d)$. Let $C_{\mathcal{F}} = \sup\{\|f\|_{ond} : f \in \mathcal{F}\}$. Set $\theta = d/(d + 2(s - d/p))$. Let Z_n be defined by (1.37). Then there exists some positive constant κ such that*

$$\|\sup_{\substack{(f,g) \in \mathcal{F} \times \mathcal{F} \\ \|f-g\|_2 \le \varepsilon}} Z_n(f - g)\|_2 \le \kappa \, C_{\mathcal{F}}^\theta \, \varepsilon^{1-\theta} \text{ for any } \varepsilon \in \,]0, C_{\mathcal{F}}]. \tag{a}$$

Consequently, if $\mathcal{F}$ is equipped with the usual norm of $L^2(\mathbb{R}^d)$, then the empirical process $\{Z_n(f) : f \in \mathcal{F}\}$ satisfies the stochastic equicontinuity condition (ii) of Theorem 7.1.

Remark 8.2 In the case $p = 2$, Theorem 8.1 holds true under the condition $s > d/2$. In that case $\theta = d/(2s)$ and $1 - \theta = (2s - d)/(2s)$. For example, if $s = d$, then $\theta = 1 - \theta = 1/2$.

Proof of Theorem 8.1. Clearly

$$\|\sup_{\substack{(f,g) \in \mathcal{F} \times \mathcal{F} \\ \|f-g\|_2 \le \varepsilon}} Z_n(f - g)\|_2 \le 2\|\sup_{\substack{f \in \mathcal{F} \\ \|f\|_2 \le \varepsilon}} |Z_n(f)| \|_2. \tag{8.3}$$

Next, by the Schwarz inequality,

$$|Z_n(f)| \le \left|\sum_{\lambda \in \mathbb{Z}^d} a_\lambda Z_n(\psi_\lambda)\right| + \sum_{j=0}^\infty \left|\sum_{\lambda \in \Lambda_j} a_\lambda Z_n(\psi_\lambda)\right|$$

$$\le \left(\sum_{\lambda \in \mathbb{Z}^d} a_\lambda^2\right)^{1/2} \left(\sum_{\lambda \in \mathbb{Z}^d} Z_n^2(\psi_\lambda)\right)^{1/2} + \sum_{j=0}^\infty \left(\sum_{\lambda \in \Lambda_j} a_\lambda^2\right)^{1/2} \left(\sum_{\lambda \in \Lambda_j} Z_n^2(\psi_\lambda)\right)^{1/2}. \tag{8.4}$$

Equations (8.2), (8.4), Lemma 8.1 and Proposition 8.1 together with the orthonormality of the wavelet basis imply that

$$\| \sup_{\substack{f \in \mathcal{F} \\ \|f\|_2 \le \varepsilon}} |Z_n(f)| \|_2 \le \varepsilon \Big(\sum_{\lambda \in \mathbb{Z}^d} \mathbb{E}(Z_n^2(\psi_\lambda)) \Big)^{1/2}$$

$$+ \sum_{j=0}^{\infty} \inf(C_{\mathcal{F}} 2^{j(d/p-s)}, \varepsilon 2^{jd/2}) \Big(\sum_{\lambda \in \Lambda_j} 2^{-jd} \mathbb{E}(Z_n^2(\psi_\lambda)) \Big)^{1/2}.$$

$$(8.5)$$

We now adapt the symmetrization method introduced in the proof of Theorem 1.3. Let $(\varepsilon_\lambda)_{\lambda \in \Lambda}$ be a sequence of independent symmetric signs, independent of the sequence $(X_i)_{i \in \mathbb{Z}}$. Then

$$\sum_{\lambda \in \Lambda_j} 2^{-jd} \mathbb{E}(Z_n^2(\psi_\lambda)) = \mathbb{E}\Big(Z_n^2 \Big(\sum_{\lambda \in \Lambda_j} 2^{-jd/2} \varepsilon_\lambda \psi_\lambda \Big) \Big). \qquad (8.6)$$

We now fix the values of the signs (ε_λ). Since the father function and the scaling functions have compact supports, there exists a positive constant K such that

$$\sum_{\lambda \in \Lambda_j} 2^{-jd/2} |\psi_\lambda(x)| \le K$$

for any x in $\mathbb{R}^d$. Consequently, if P denotes the law of X_0, then for any family (ε_λ) of signs, the quantile function of the random variable

$$\Big(\sum_{\lambda \in \Lambda_j} 2^{-jd/2} \varepsilon_\lambda \psi_\lambda(X_0) \Big)^2$$

is bounded above by K^2. Hence, both (8.6) and Corollary 1.2(b) ensure that

$$\sum_{\lambda \in \Lambda_j} 2^{-jd} \mathbb{E}(Z_n^2(\psi_\lambda)) \le 4K^2 \sum_{k=0}^{n-1} \alpha_k.$$

The same upper bound holds true for the scale V_0, whence

$$\Big(\mathbb{E}\Big(\sup_{\substack{f \in \mathcal{F} \\ \|f\|_2 \le \varepsilon}} Z_n^2(f) \Big) \Big)^{1/2} \le 2K \Big(\sum_{k=0}^{n-1} \alpha_k \Big)^{1/2} \Big(\varepsilon + \sum_{j=0}^{\infty} \inf(C_{\mathcal{F}} 2^{j(d/p-s)}, \varepsilon 2^{jd/2}) \Big).$$

$$(8.7)$$

Suppose now that $\varepsilon \le C_{\mathcal{F}}$. Let r be the nonnegative real such that $C 2^{r(d/p-s)} = \varepsilon 2^{rd/2}$. Then

$$\sum_{j=0}^{\infty}\inf(C_{\mathcal{F}}2^{j(d/p-s)}, \varepsilon 2^{jd/2}) = \sum_{j\leq r}\varepsilon 2^{jd/2} + \sum_{j>r}C_{\mathcal{F}}2^{j(d/p-s)}$$

$$\leq \varepsilon 2^{rd/2}(1 - 2^{-d/2})^{-1} + C_{\mathcal{F}}2^{r(d/p-s)}(1 - 2^{d/p-s})^{-1}.$$
$$(8.8)$$

Next, by definition of r,

$$C_{\mathcal{F}}2^{r(d/p-s)} = \varepsilon 2^{rd/2} = C_{\mathcal{F}}^{1/(1+2(s/d)-2/p)}\varepsilon^{(2(s/d)-2/p)/(1+2(s/d)-2/p)} = C_{\mathcal{F}}^{\theta}\varepsilon^{1-\theta}.$$

Let $C(d, p) = 1 + (1 - 2^{-d/2})^{-1} + (1 - 2^{d/p-s})^{-1}$. Both (8.7) and (8.8) together with the above equalities ensure that

$$\left(\mathbb{E}\Big(\sup_{\substack{f\in\mathcal{F}\\ \|f\|_2\leq\varepsilon}} Z_n^2(f)\Big)\right)^{1/2} \leq 2KC(d, p)\Big(\sum_{k=0}^{n-1}\alpha_k\Big)^{1/2}C_{\mathcal{F}}^{\theta}\varepsilon^{1-\theta},\qquad(8.9)$$

provided that $\varepsilon \leq C$. Theorem 8.1 then follows from (8.3) and (8.9). ∎

From Theorem 8.1, Corollary 4.1 and Theorem 7.1 (Theorem 10.2 in Pollard 1990), we now derive the following uniform central limit theorem.

Corollary 8.1 *Let $(X_i)_{i\in\mathbb{Z}}$ be a stationary and ergodic sequence of real-valued random variables with values in $\mathbb{R}^d$. Suppose that the strong mixing coefficients defined by (2.1) satisfy $\sum_{k\geq 0}\alpha_k < \infty$. Let p be a real in $[1, 2]$ and let $s > d/p$. Let $\mathcal{F}$ be a bounded and closed subset of $\mathrm{Lip}^*(s, p, \mathbb{R}^d)$. Suppose furthermore that $\mathcal{F}$ is totally bounded in $L^2(\mathbb{R}^d)$ (see Definition 8.4). Then there exists a Gaussian process G with a.s. uniformly continuous trajectories on the space $\mathcal{F}$ equipped with the usual metric in $L^2(\mathbb{R}^d)$, such that $\{Z_n(f) : f \in \mathcal{F}\}$ converges in distribution to G in the sense of Theorem 7.1.*

8.3   Maximal Coupling and Entropy with Bracketing

Throughout this section, $(X_i)_{i\in\mathbb{Z}}$ is a strictly stationary and absolutely regular sequence of random variables with values in some Polish space $\mathcal{X}$. The absolute regularity or β-mixing coefficients are defined as in Sect. 5.5. Below we recall the definition of these coefficients.

Definition 8.3 The absolute regularity or β-mixing coefficients $(\beta_n)_{n\geq 0}$ of the sequence $(X_i)_{i\in\mathbb{Z}}$ are defined by $\beta_0 = 1$ and

$$\beta_n = \sup_{k\in\mathbb{Z}}\beta(\mathcal{F}_k, \mathcal{G}_{k+n}) \text{ for } n > 0,\qquad(8.10)$$

with the same notations as in Definitions 8.2 and 8.1.

Throughout the sequel, we denote by P the law of X_0. Z_n denotes the normalized empirical measure, as defined in (1.37). We will assume that the sequence of β-mixing coefficients satisfies the summability condition $\sum_{n>0} \beta_n < \infty$. Our aim is to extend the uniform central limit theorem of Dudley (1978) for empirical processes indexed by classes of function with an integrable $L^1(P)$-entropy with bracketing to the β-mixing case. Using the maximal coupling lemma of Goldstein (1979) or Berbee (1979), we will construct a positive measure Q with finite total mass, absolutely continuous with respect to P with the following remarkable property: for any class $\mathcal{F}$ of uniformly bounded functions with an integrable $L^1(Q)$-entropy with bracketing, the empirical process $\{Z_n(f) : f \in \mathcal{F}\}$ satisfies the uniform functional central limit theorem.

We now recall the definitions of metric entropy and of metric entropy with bracketing and Dudley (1978) functional central limit theorem for empirical processes associated to strictly stationary sequences of independent random variables. We start with the definition of metric entropy.

Definition 8.4 Let (V, d) be a pseudo-metric space. Let

$$N(\delta, V, d) = \min\{n \in \mathbb{N} : \exists S_n = \{x_1, \ldots, x_n\} \subset S \text{ such that } d(x, S_n) \leq \delta \text{ for any } x \in S\}.$$

(V, d) is said to be totally bounded if $N(\delta, V, d) < \infty$ for any positive δ. The metric entropy function H is defined by $H(\delta, V, d) = \log(N(\delta, V, d) \vee 2)$.

Dudley (1967) has given an entropy criterion ensuring the a.s. uniform continuity of Gaussian processes. Let H be a Hilbert space and B be a Gaussian process indexed by H, with covariance function the scalar product of H. If V is a totally bounded subset of H and if

$$\int_0^1 \sqrt{H(x, V, d)}\, dx < \infty, \tag{8.11}$$

then there exists a version of B with a.s. uniformly continuous trajectories on V.

However, as shown by some counterexamples, condition (8.11) does not imply the uniform functional central limit theorem for empirical processes. Some additional conditions are needed, such as bracketing conditions. Below we define the notions of brackets and diameter of brackets and the notion of entropy with bracketing.

Definition 8.5 Let V be a subspace of the space of numerical functions on $(\mathcal{X}, P)$. Let $\Lambda : V \to \mathbb{R}^+$ be a function such that, for any f and any g in V,

$$|f| \leq |g| \text{ implies that } \Lambda(f) \leq \Lambda(g). \tag{8.12}$$

Let $\mathcal{F} \subset V$. If $f \leq g$, we denote by $[f, g]$ the set of functions h such that $f \leq h \leq g$. This set is called an interval of functions. The nonnegative real $\Lambda(g - f)$ is called the diameter of $[f, g]$.

A class $\mathcal{F}$ of functions in V is said to be totally bounded with bracketing if, for any positive δ, there exists a finite family $S(\delta)$ of intervals of functions in V

with diameter less than δ, such that

$$\text{for any } f \in \mathcal{F}, \text{ there exists } [g, h] \in S(\delta) \text{ such that } f \in [g, h]. \qquad (8.13a)$$

The covering number $\mathcal{N}_{[\,]}(\delta, \mathcal{F})$ in (V, Λ) is the minimal cardinality of families $S(\delta)$ satisfying (8.13a). The entropy with bracketing is defined by

$$H_{[\,]}(\delta, \mathcal{F}, \Lambda) = \log \mathcal{N}_{[\,]}(\delta, \mathcal{F}) \vee 2. \qquad (8.13b)$$

If Λ is a norm on V and if d_λ is the distance corresponding to this norm, then the following relation between entropy and entropy with bracketing holds:

$$H(\delta, \mathcal{F}, d_\Lambda) \leq H_{[\,]}(2\delta, \mathcal{F}, \Lambda). \qquad (8.14)$$

In the general case, the converse inequality does not hold. The notions of entropy and entropy with bracketing are not equivalent. The only notable exception is the case of the uniform distance, which corresponds to

$$\Lambda(f) = \|f\|_\infty = \sup_{x \in \mathcal{X}} |f(x)|. \qquad (8.15)$$

In that case $\bar{B}(f, \delta) = [f - \delta, f + \delta]$, and consequently balls are intervals. Then

$$H_{[\,]}(2\delta, \mathcal{F}, \|\cdot\|_\infty) = H(\delta, \mathcal{F}, \|\cdot\|_\infty). \qquad (8.16)$$

We now recall Ossiander's theorem (1987) for empirical processes indexed by classes of functions. This result is an extension of Dudley (1978) Theorem to $L^2(P)$. Let $(X_i)_{i \in \mathbb{Z}}$ be a strictly stationary sequence of independent random variables and let P denote the law of X_0. Throughout the sequel, let the normalized empirical measures Z_n be defined by (1.37). In the iid case, for any f in $L^2(P)$,

$$\text{Var} Z_n(f) = \int f^2 dP - \left(\int f dP \right)^2. \qquad (8.17)$$

Consequently, if $\mathcal{F}$ is a class of functions in $L^2(P)$, then the fidi convergence of $\{Z_n(f) : f \in \mathcal{F}\}$ to an a.s. uniformly continuous Gaussian process G with covariance function $\Gamma(f, g) = \int f g dP - \int f dP \int g dP$ holds as soon as Dudley's criterion is satisfied, i.e.

$$\int_0^1 \sqrt{H(x, \mathcal{F}, d_P)} \, dx < \infty, \qquad (8.18)$$

where d_P is defined by

$$d_P^2(f, g) = \int (f - g)^2 dP - \left(\int (f - g) dP \right)^2. \qquad (8.19)$$

Condition (8.18) does not imply the asymptotic stochastic equicontinuity of Z_n. However, the corresponding bracketing condition implies the functional central limit theorem for the empirical process.

Theorem 8.2 (Ossiander (1987)) *Let $(X_i)_{i \in \mathbb{Z}}$ be a sequence of independent random variables with common law P and let $\mathcal{F} \subset L^2(P)$. If $\mathcal{F}$ is totally bounded with bracketings in $L^2(P)$ and if*

$$\int_0^1 \sqrt{H_{[]}(x, \mathcal{F}, d_{2,P})}\, dx < \infty, \tag{8.20}$$

then the empirical process $\{Z_n(f) : f \in \mathcal{F}\}$ satisfies the uniform functional central limit theorem.

Ossiander's theorem is a remarkable extension of Dudley (1978) theorem for classes of uniformly bounded functions with integrable entropy with bracketing in $L^1(P)$. We refer to Andersen et al. (1988) for more precise results in the independent case. Doukhan et al. (1995) give the following extension of Ossiander's theorem to β-mixing sequences. For any numerical function f, we set, for the sake of convenience, $Q_f = Q_{f(X_0)}$. Let us define the norm $\| \cdot \|_{2,\beta}$ by

$$\|f\|_{2,\beta} = \left(\int_0^1 \beta^{-1}(u) Q_f^2(u) du \right)^{1/2}. \tag{8.21}$$

This norm satisfies (8.12). Hence we may consider the entropy with bracketing with respect to this new norm. Let $L_{2,\beta}(P)$ be the space of functions f such that $\|f\|_{2,\beta} < \infty$. (Doukhan et al., 1995) prove that the uniform functional central limit theorem holds for a class of functions $\mathcal{F}$ included in $L_{2,\beta}(P)$ as soon as

$$\int_0^1 \sqrt{H_{[]}(x, \mathcal{F}, \| \cdot \|_{2,\beta})}\, dx < \infty. \tag{8.22}$$

Let us now apply this result to classes of uniformly bounded functions. Let $\mathcal{F}$ be a class of numerical functions with values in $[-1, 1]$. Then the fidi convergence to a Gaussian process holds as soon as the summability condition $\sum_{n>0} \beta_n < \infty$ is satisfied. In the general case, (8.22) needs a stronger mixing condition. The same gap appears in the paper by Arcones and Yu (1994). Nevertheless, since

$$\|f\|_{2,\beta} \le \|f\|_\infty \sum_{n \ge 0} \beta_n,$$

Equation (8.22) implies the functional uniform central limit theorem under the minimal mixing condition $\sum_{n>0} \beta_n < \infty$ if $\mathcal{F}$ satisfies the stronger entropy condition

$$\int_0^1 \sqrt{H(x, \mathcal{F}, \| \cdot \|_\infty)}\, dx < \infty. \tag{8.23}$$

However, Condition (8.23) is not relevant for classes of sets. If $\mathcal{F}$ is the class of indicator functions of orthants or Euclidean balls, (8.22) needs the mixing condition

$$\sum_{n \geq 2} n^{-1} (\log n)^{-1/2} \Big(\sum_{i \geq n} \beta_i \Big)^{1/2} < \infty. \tag{8.24}$$

For example, if $\beta_n = O(n^{-1}(\log n)^{-b})$, (8.23) needs the too restrictive condition $b > 2$. By contrast, Rio (1998) obtains the uniform functional central limit theorem for these classes of sets under the minimal mixing condition $\sum_{n>0} \beta_n < \infty$. His approach is based on repeated application of the maximal coupling lemma. Here we will adapt Rio's approach to the case of entropy with bracketing. As in Rio (1998), we construct a nonnegative function B in $L^1(P)$ such that the positive measure $Q = BP$ has some nice properties, which will allow us to prove the stochastic equicontinuity of the empirical process as soon as the class of functions $\mathcal{F}$ has an integrable entropy with bracketing in $L^1(Q)$.

Before stating the main result of this section, we need to define Q. In order to define Q, we will use the maximal coupling theorem (see Theorem 5.1). Applying Theorem 5.1 to $(X_i)_{i \in \mathbb{Z}}$, we get that there exists some sequence $(X_i^*)_{i \in \mathbb{Z}}$ of random variables with the following properties: the sequence $(X_i^*)_{i \in \mathbb{Z}}$ has the same law as $(X_i)_{i \in \mathbb{Z}}$, is independent on $\mathcal{F}_0 = \sigma(X_i : i \leq 0)$ and

$$\mathbb{P}(X_i = X_i^* \text{ for any } i \geq k) = 1 - \beta_k.$$

We now define Q from the above coupling sequence. Since X_k^* is independent of X_k,

$$\mathrm{Cov}(f(X_0), f(X_k)) = \mathbb{E}\big(f(X_0)(f(X_k) - f(X_k^*))\big) \tag{8.25}$$

for any bounded function f. Consequently,

$$|\mathrm{Cov}(f(X_0), f(X_k))| \leq 2\|f\|_\infty \mathbb{E}(|f(X_k) - f(X_k^*)|). \tag{8.26}$$

Now

$$\mathbb{E}(|f(X_k) - f(X_k^*)|) \leq \mathbb{E}(|f(X_k)| \mathrm{I\!I}_{X_k \neq X_k^*}) + \mathbb{E}(|f(X_k^*)| \mathrm{I\!I}_{X_k \neq X_k^*}). \tag{8.27}$$

Let us then define the measurable functions b_k' and b_k^* from $\mathcal{X}$ into $[0, 1]$ by

$$b_k'(X_k) = \mathbb{P}(X_k \neq X_k^* \mid X_i) \text{ and } b_k^*(X_k^*) = \mathbb{P}(X_k \neq X_k^* \mid X_k^*). \tag{8.28}$$

From (8.27) we get that

$$\mathbb{E}(|f(X_k) - f(X_k^*)|) \leq \mathbb{E}(|f(X_k)|b_k'(X_k)) + \mathbb{E}(|f(X_k^*)|b_k^*(X_k^*)). \tag{8.29}$$

Hence, if $b_k = (b_k' + b_k^*)/2$,

$$\mathbb{E}(|f(X_k) - f(X_k^*)|) \le 2 \int_{\mathcal{X}} b_k |f| dP, \tag{8.30}$$

which, together with (8.26), implies

$$\operatorname{Var} Z_n(f) \le \|f\|_\infty \int_{\mathcal{X}} (1 + 4b_1 + \cdots + 4b_{n-1}) |f| dP. \tag{8.31}$$

We then define the nonnegative measure Q by

$$Q = BP = \left(1 + 4 \sum_{k>0} b_k\right) P. \tag{8.32a}$$

From (8.25) and (8.28), the functions b_k satisfy the additional conditions

$$0 \le b_k \le 1 \text{ and } \int_{\mathcal{X}} b_k dP \le \beta_k. \tag{8.32b}$$

Consequently, under the summability condition, the measure Q has a finite total mass and is absolutely continuous with respect to P. Furthermore, the following uniform functional central limit theorem holds for the measure Q defined by (8.32a).

Theorem 8.3 *Let $(X_i)_{i \in \mathbb{Z}}$ be a strictly stationary sequence of random variables with values in some Polish space $\mathcal{X}$, with common law P, and let Q be the nonnegative measure defined by (8.32a). Suppose that the sequence $(\beta_k)_{k>0}$ of absolute regularity coefficients defined by (8.10) satisfies the summability condition $\sum_{k>0} \beta_k < \infty$. Let $\mathcal{F}$ be a class of measurable functions from $\mathcal{X}$ into $[0, 1]$. Let*

$$\|f\|_{1,Q} = \int_{\mathcal{X}} |f| dQ \text{ and } d_{1,Q}(f, g) = \|f - g\|_{1,Q}.$$

Let $L^1(Q)$ be the space of numerical functions f such that $\|f\|_{1,Q} < \infty$. Suppose that $\mathcal{F}$ is totally bounded with bracketings in $L^1(Q)$ and that

$$\int_0^1 \sqrt{H_{[]}(x, \mathcal{F}, d_{1,Q})/x}\, dx < \infty. \tag{8.33}$$

Then the empirical process $\{Z_n(f) : f \in \mathcal{F}\}$ satisfies the uniform functional central limit theorem of Theorem 7.1.

Applications of Theorem 8.3. Let us first note that Theorem 8.3 is not adequate for classes of regular functions satisfying entropy conditions in $L^\infty(P)$. Indeed, condition (8.23) does not imply (8.37). The main interest of Theorem 8.3 lies in the. Suppose that $\mathcal{F} = \{\mathbb{I}_S : S \in \mathcal{S}\}$. Since

$$\|\mathbb{I}_S - \mathbb{I}_T\|_{1,Q} = Q(S \Delta T) \tag{8.34}$$

(here Δ denotes the symmetric difference), the notions of entropy with bracketing and entropy with inclusion are equivalent (see Dudley (1978) for a definition of entropy with inclusion and a detailed exposition). In that case condition (8.33) is equivalent to the following condition of entropy with bracketings in $L^2(Q)$:

$$\int_0^1 \sqrt{H_{[]}(x, \mathcal{F}, d_{2,Q})}\, dx < \infty \quad \text{with } d_{2,Q}(f, g) = \left(\int_{\mathcal{X}} (f - g)^2 dQ\right)^{1/2}. \quad (8.35)$$

Since $\|f\|_{2,Q} \leq 2\|f\|_{2,\beta}$, (8.35) is weaker than Doukhan, Massart and Rio's entropy condition (8.22). Generally (8.35) provides better results for classes of sets. For example, for the class of orthants, the uniform central limit theorem holds under the minimal mixing condition $\sum_{k>0} \beta_k < \infty$, as shown by Corollary 8.2 below.

Corollary 8.2 *Let $(X_i)_{i \in \mathbb{Z}}$ be a strictly stationary sequence of random variables with values in $\mathbb{R}^d$, satisfying the β-mixing condition $\sum_{k>0} \beta_k < \infty$. Assume that the marginal distribution functions F_j of the law of X_0 are continuous. Then there exists a Gaussian process G with uniformly continuous trajectories on $\mathbb{R}^d$ equipped with the pseudo-metric d_F given by*

$$d_F(x, y) = \sup_{j \in [1,d]} |F_j(x_j) - F_j(y_j)|, \text{ where } x = (x_1, \ldots x_d) \text{ and } y = (y_1, \ldots y_d),$$

such that the normalized and centered empirical distribution function ν_n defined in Sect. 7.5 converges in distribution to G in $B(\mathbb{R}^d)$ as n tends to ∞.

Proof of Theorem 8.3. We start by replacing the initial entropy function by some function H with additional monotonicity properties.

Lemma 8.2 *Let $H_{1,Q}(x) = H_{[]}(x, \mathcal{F}, d_{1,Q})$. There exists some continuous and non-increasing function $H \geq H_{1,Q}$ such that the function $x \to x^2 H(x)$ is nondecreasing and*

$$\int_0^v (H(x)/x)^{1/2} dx \leq 2 \int_0^v (H_{1,Q}(x)/x)^{1/2} dx \text{ for any } v \in]0, 1]. \quad (8.36)$$

Proof Let $H(x) = \sup_{t \in]0,x]} (t/x)^2 H_{1,Q}(t)$. By definition $H \geq H_{1,Q}$, H is nonincreasing and $x \to x^2 H(x)$ is nondecreasing, which implies that H is continuous. Next

$$x\sqrt{H(x)} \leq \sup_{t \in]0,x]} t\sqrt{H_{1,Q}(t)} \leq \int_0^x \sqrt{H_{1,Q}(t)}\, dt,$$

whence

$$\int_0^v (H(x)/x)^{1/2} dx \leq \int_0^v \int_0^v \mathbb{I}_{t<x} \sqrt{H_{1,Q}(t)} x^{-3/2} dt dx \leq 2 \int_0^v (H_{1,Q}(x)/x)^{1/2} dx$$

by Fubini's Theorem. Hence Lemma 8.2 holds true. ∎

We now prove Theorem 8.3: the main step is to prove the stochastic equicontinuity property. If the function H is uniformly bounded, then S is finite. In that case the uniform central limit theorem follows directly from the fidi convergence. Consequently, we may assume that $\lim_0 H(x) = +\infty$. We start with some definitions.

Definition 8.6 Let δ be a fixed positive real. Let K be the first nonnegative integer such that $2^{2K} H(\delta) \geq n\delta$. Set $q_0 = 2^K$. For any integer k in $[1, K]$, let $q_k = q_0 2^{-k}$ and let δ_k be the unique positive real satisfying $q_k^2 H(\delta_k) = n\delta_k$. Let $\delta_0 = \delta$.

The first step of the proof is to replace the initial sequence by a sequence of independent blocks of length q_0. This will be done using the coupling Lemma 5.1. Applying Lemma 5.1 recursively, we get that there exists some sequence $(X_i^0)_{i>0}$ with the properties below.

1. Let $q = q_0$. For any $i \geq 0$, the random vector $U_i^0 = (X_{iq+1}^0, \ldots, X_{iq+q}^0)$ has the same law as $U_i = (X_{iq+1}, \ldots, X_{iq+q})$.
2. The random vectors $(U_{2i}^0)_{i\geq 0}$ form an independent sequence, and the same property holds for $(U_{2i+1}^0)_{i\geq 0}$.
3. Furthermore, $\mathbb{P}(U_i \neq U_i^0) \leq \beta_q$ for any $i \geq 0$.

Using properties 1–3, we now bound the cost of replacement of the initial sequence by this new sequence.

Lemma 8.3 Let $S_n^0(f) = f(X_1^0) + \cdots + f(X_n^0)$ and $Z_n^0(f) = n^{-1/2}(S_n^0(f) - n P(f))$. Set $q = q_0$. Then

$$\mathbb{E}^*\left(\sup_{f\in\mathcal{F}} |Z_n(f) - Z_n^0(f)|\right) \leq 2\sqrt{n}\,\beta_q.$$

Proof Set $S_n(f) = f(X_1) + \cdots + f(X_n)$. For any f in $\mathcal{F}$,

$$|S_n(f) - S_n^0(f)| \leq \sum_{i=1}^n |f(X_i) - f(X_i^0)| \leq 2\sum_{i=1}^n \mathbb{I}_{X_i\neq X_i^0}. \tag{8.37}$$

Hence, by Property 3,

$$\mathbb{E}^*\left(\sup_{f\in\mathcal{F}} |Z_n(f) - Z_n^0(f)|\right) \leq 2n^{-1/2}\sum_{i=1}^n \mathbb{P}(X_i \neq X_i^0) \leq 2\sqrt{n}\,\beta_q, \tag{8.38}$$

which completes the proof of Lemma 8.3. ∎

Now, from Definition 8.6, $q_0^2 \geq n\delta/H(\delta)$. Since $\lim_{n\uparrow\infty} n\beta_n = 0$, this ensures that $\lim_{n\uparrow\infty} \sqrt{n}\beta_{q_0} = 0$. Consequently, the upper bound in Lemma 8.3 tends to 0 as n tends to ∞. Thus the stochastic equicontinuity property holds for Z_n if and only if this property holds true for Z_n^0.

Let us now prove the stochastic equicontinuity property for Z_n^0. The main problem which arises here is that the length of blocks q_0 is too large to get efficient

Bernstein type exponential inequalities. In order to improve the results of Doukhan et al. (1995) or Arcones and Yu (1994), we will recursively replace the sequence $(X_n^0)_n$ by sequences $(X_n^j)_n$ with independent blocks of length $q_j = q_0 2^{-j}$. In order to construct the sequences $(X_n^j)_n$, we will assume that the underlying probability space is rich enough in the following sense: there exists an array $(u_{j,i})_{(j,i)\in\mathbb{N}\times\mathbb{N}}$ of independent random variables with uniform law on $[0, 1]$, independent of the sequence $(X_i^0)_i$.

We first construct the sequence $(X_l^1)_l$ from $(X_l^0)_l$. Let $q = q_0$. Then $q_1 = q/2$ and

$$W_i^0 = (U_i^0, V_i^0) \text{ with } U_i^0 = (X_{qi+1}^0, \ldots X_{qi+q_1}^0) \text{ and } V_i^0 = (X_{qi+q_1+1}^0, \ldots X_{qi+q_0}^0).$$

By the maximal coupling theorem (Theorem 5.1) applied to the sequence $(\xi_k)_{-q_1 < k \le q_1}$ defined by $\xi_k = X_{k-qi-q_1}$ together with Skorohod's lemma (Lemma E.2), there exists a random sequence $(\xi_k^*)_{-q_1 < k \le q_1}$ with the same distribution as $(\xi_k)_{-q_1 < k \le q_1}$, which is a measurable function of $u_{0,i}$ and W_i^0, such that the random vector $(\xi_1^*, \ldots, \xi_{q_1}^*)$ is independent of U_i^0 and has the same distribution as V_i^0. We then set (here $q = q_0$)

$$X_l^1 = X_l^0 \text{ for } l \in [iq + 1, iq + q_1] \text{ and } X_l^1 = \xi_{l+qi+q_1}^* \text{ for } l \in]iq + q_1, iq + q]. \tag{8.39}$$

From (8.39) together with (8.30) and (8.32), for any bounded function f,

$$\sum_{l=iq_0+1}^{iq_0+q_0} \mathbb{E}(|f(X_i^0) - f(X_i^1)|) \le \|f\|_{1,Q}.$$

Proceeding by induction one can prove the proposition below.

Proposition 8.2 *Let $(X_i^*)_{i\in\mathbb{Z}}$ be defined from $(X_i)_{i\in\mathbb{Z}}$ via Theorem 5.1. Then one can construct sequences $(X_i^j)_{i>0}$ for j in $[1, K]$ with the properties below:*

(i) Let $q = q_0$. Set $T_i^j = (X_{iq+1}^j, \ldots, X_{iq+q}^j)_{j\in[0,K]}$. Then the blocks $(T_i^j)_{i\ge 0}$ are identically distributed. Furthermore, the blocks $(T_{2i}^j)_{i\ge 0}$ are mutually independent and the blocks $(T_{2i+1}^j)_{i\ge 0}$ are mutually independent.
(ii) For j in $[0, K]$, let

$$W_i^j = (U_i^j, V_i^j) \text{ with } U_i^j = (X_{q_j i+1}^j, \ldots X_{q_j i+q_{j+1}}^j) \text{ and } V_i^j = (X_{q_j i+q_{j+1}+1}^j, \ldots X_{q_j i+q_j}^j).$$

Then, for any j in $[1, K]$ and any nonnegative integer i, $W_{2i}^j = U_i^{j-1}$, W_{2i+1}^j is a measurable function of W_i^{j-1} and $u_{j-1,i}$, and the random vector (W_i^{j-1}, W_{2i+1}^j) has the same distribution as $(X_{1-q_j}, \ldots, X_{q_j}, X_1^*, \ldots X_{q_j}^*)$.
Properties (i) and (ii) ensure the following additional properties:
(iii) For any j in $[0, K]$, the random vectors $W_0^j, \ldots, W_{2^j-1}^j$ are independent and identically distributed, and W_0^j has the same law as $(X_1, \ldots, X_{q_j})$.
(iv) For any bounded function f, any j in $[0, K-1]$ and any $i \ge 0$,

$$\sum_{l=iq_j+1}^{iq_j+q_j} \mathbb{E}(|f(X_i^j - f(X_i^{j+1})|) \leq \|f\|_{1,Q}.$$

In order to control the fluctuations, we will define δ_k-nets $\mathcal{F}_k$ as well as projections Π_k on $\mathcal{F}_k$.

Definition 8.7 For any k in $[0, K]$, let $\mathcal{J}_k = \{B_{1,k}, B_{2,k}, \ldots\}$ be a totally ordered collection of intervals of functions with diameter less than δ_k with respect to $\|\cdot\|_{1,Q}$, such that $\mathcal{F} \subset \bigcup_{[g,h]\in\mathcal{J}_k}[f, g]$ and $\log \operatorname{Card} \mathcal{F}_k \leq H(\delta_k)$. For each interval $B_{j,k} = [g_{j,k}, h_{j,k}]$ in $\mathcal{J}_k$, we choose a point $f_{j,k}$ in $B_{j,k} \cap \mathcal{F}$. For any f in $\mathcal{F}$, let j be the first integer such that f belongs to $B_{j,k}$. We set $\Pi_k f = f_{j,k}$ and $\Delta_k f = h_{j,k} - g_{j,k}$. We denote by $\mathcal{F}_k$ the set of functions $\Pi_k f$ as f ranges over $\mathcal{F}$ and by $\mathcal{G}_k$ the set of functions $\Delta_k f$.

From Definition 8.7, the operators Π_k and Δ_k satisfy

$$|f - \Pi_k f| \leq \Delta_k f \quad \text{and} \quad \|\Delta_k f\|_{1,Q} \leq \delta_k, \ \|\Delta_k f\|_\infty \leq 2. \tag{8.40}$$

We now introduce our chaining argument. Since this chaining argument has to be adapted to the dependence setting, the above defined sequences will play a fundamental role. Here we need to introduce some additional notation.

Notation 8.1. Let $S_n^k(f) = f(X_1^k) + \cdots + f(X_n^k)$ and $Z_n^k(f) = n^{-1/2}(S_n^k(f) - n P(f))$.

We now give our chaining decomposition:

$$Z_n^0(f - \Pi_0 f) = Z_n^K(f - \Pi_K f) + \sum_{l=1}^{K} Z_n^{l-1}(\Pi_l f - \Pi_{l-1} f)$$

$$+ \sum_{k=1}^{K}(Z_n^{k-1} - Z_n^k)(f - \Pi_k f). \tag{8.41}$$

From the decomposition (8.41),

$$\mathbb{E}^*(\sup_{f\in\mathcal{F}} |Z_n^0(f - \Pi_0 f)|) \leq \mathbb{E}_1 + \mathbb{E}_2 + \mathbb{E}_3, \tag{8.42}$$

with

$$\mathbb{E}_1 = \mathbb{E}(\sup_{f\in\mathcal{F}} |Z_n^K(f - \Pi_K f)|),$$

$$\mathbb{E}_2 = \sum_{l=1}^{K} \mathbb{E}(\sup_{f\in\mathcal{F}}(|Z_n^{l-1}(\Pi_l f - \Pi_{l-1} f)|),$$

$$\mathbb{E}_3 = \sum_{k=1}^{K} \mathbb{E}(\sup_{f\in\mathcal{F}}(|(Z_n^{k-1} - Z_n^k)(f - \Pi_k f)|).$$

Control of $\mathbb{E}_1$

Since the random variables X_i^K have common law P, by (8.40),

$$
\begin{aligned}
|Z_n^K(f - \Pi_K f)| &\le n^{-1/2}(S_n^K(|f - \Pi_K f|) + nP(|f - \Pi_K f|)) \\
&\le n^{-1/2}(S_n^K(\Delta_K f) + nP(\Delta_K f)) \\
&\le Z_n^K(\Delta_K f) + 2\sqrt{n}P(\Delta_k f).
\end{aligned}
\tag{8.43}
$$

Now $P \le Q$ and $Q(\Delta_K f) \le \delta_K$, whence

$$
\mathbb{E}_1 \le 2n^{1/2}\delta_K + \mathbb{E}(\sup_{g \in \mathcal{G}_K} Z_n^K(g)).
\tag{8.44}
$$

By Proposition 8.2(ii), the random vectors (X_{2i+1}^k, X_{2i+2}^k) have the same law as (X_0, X_1^*). Consequently, by Proposition 8.2(iii), the random variables $X_1^K, \ldots, X_{q_0}^K$ are independent and with common law P. Next

$$
S_n^K(g) = A + B \text{ with } A = \sum_{\substack{(i-1)/q_0 \in 2\mathbb{N} \\ i \le n}} g(X_i^K) \text{ and } B = \sum_{\substack{(i-1)/q_0 \in 2\mathbb{N}+1 \\ i \le n}} g(X_i^K).
$$

Now, by the Schwarz inequality,

$$
\log \mathbb{E}\big(\exp t(S_n^K(g) - nP(g))\big) \le \frac{1}{2}\Big(\log \mathbb{E}(\exp(2t(A - \mathbb{E}(A)) + \log \mathbb{E}(\exp(2t(B - \mathbb{E}(B))\big).
$$

By Proposition 8.2(i) (and (iii)), A and B are sums of independent random variables with the same law as $g(X_1)$. Hence, applying Inequality (B.4) in Annex B to A and B, we obtain that

$$
\log \mathbb{E}\big(\exp t(S_n^K(g) - nP(g))\big) \le \frac{nP(g^2)t^2}{1 - 4t/3} \le \frac{n\|g\|_{1,Q}t^2}{1 - 4t/3}
\tag{8.45}
$$

for any g with $\|g\|_\infty \le 1$. Since $\|g\|_\infty \le 1$ for any g in $\mathcal{G}_k$ and the logarithm of the cardinality of $\mathcal{G}_k$ is less than $H(\delta_K)$, both (8.45) and Inequality (B.5) together with Lemma D.1 in Annex D then imply that

$$
\mathbb{E}(\sup_{g \in \mathcal{G}_K} Z_n^K(g)) \le 2\sqrt{\delta_K H(\delta_K)} + 2n^{-1/2}H(\delta_K).
$$

Now, by definition of δ_k,

$$
n^{-1/2}q_k H(\delta_k) = n^{1/2}(\delta_k/q_k) = (\delta_k H(\delta_k))^{1/2}.
\tag{8.46}
$$

Since $q_K = 1$, combining (8.44) with the above inequalities, we finally get that

$$
\mathbb{E}_1 \le 6\sqrt{\delta_K H(\delta_K)}.
\tag{8.47}
$$

Control of $\mathbb{E}_2$

Fix l in $[0, K-1]$ and let

$$\mathbb{E}_{2,l} = \mathbb{E}(\sup_{f \in \mathcal{F}} (|Z_n^l(\Pi_{l+1}f - \Pi_l f)|).$$

By Proposition 8.2 (ii)–(iii), the random vectors $(X_{iq_l+1}^l, \ldots, X_{(i+1)q_l}^l)_{i \in [0, 2^l-1]}$ are independent and identically distributed. Next

$$S_n^l(g) = A_l + B_l \text{ with } A = \sum_{\substack{(i-1)/q_0 \in 2\mathbb{N} \\ i \le n}} g(X_i^l) \text{ and } B = \sum_{\substack{(i-1)/q_0 \in 2\mathbb{N}+1 \\ i \le n}} g(X_i^l).$$

Again, by Proposition 8.2(i) and (iii), A_l and B_l are sums of independent random variables with the same law as $g(X_1) + \cdots + g(X_{q_l})$, with the exception of the last block, which has the same distribution as $g(X_1) + \cdots + g(X_{n-q_l[n/q_l]})$. Suppose now that $\|g\|_\infty \le 1$. Then these independent random variables are bounded above by $q_l \|g\|_\infty$ and, by (8.31), their variance is bounded above by the length of the block multiplied by $\| \|g\|_{1,\varrho}$. Hence, as before, by the Schwarz inequality together with Inequality (B.4),

$$\log \mathbb{E}(\exp t(S_n^l(g) - nP(g))) \le n\|g\|_{1,\varrho} t^2/(1 - 4q_l t/3). \tag{8.48}$$

Let

$$\mathcal{U}_l = \{\Pi_l f - \Pi_{l+1} f : f \in \mathcal{F}\}.$$

For any g in $\mathcal{U}_l$, $\|g\|_\infty \le 1$, and $\|g\|_{1,\varrho} \le 2\delta_l$. Since the logarithms of the cardinalities of $\mathcal{U}_l$ and $-\mathcal{U}_l$ are less than $2H(\delta_{l+1})$, both (8.45) and Inequality (B.5) together with Lemma D.1 in Annex D applied successively to $\mathcal{U}_l$ and $-\mathcal{U}_l$ then imply that

$$\mathbb{E}_{2,l} \le 4\sqrt{\delta_l H(\delta_{l+1})} + 4n^{-1/2} q_l H(\delta_{l+1}). \tag{8.49}$$

Now, by definition of δ_{l+1},

$$q_l H(\delta_{l+1}) = 2(q_{l+1}^2 H(\delta_{l+1}) H(\delta_{l+1}))^{1/2} = 2(n\delta_{l+1} H(\delta_{l+1}))^{1/2},$$

whence $\mathbb{E}_{2,l} \le 12(\delta_l H(\delta_{l+1}))^{1/2}$. It follows that

$$\mathbb{E}_2 \le \sum_{l=0}^{K-1} \mathbb{E}_{2,l} \le 12 \sum_{l=0}^{K-1} \sqrt{\delta_l H(\delta_{l+1})}. \tag{8.50}$$

Control of $\mathbb{E}_3$.

For k in $[1, K]$, let

$$\mathbb{E}_{3,k} = \mathbb{E}(\sup_{f \in \mathcal{F}} (|(Z_n^{k-1} - Z_n^k)(f - \Pi_k f)|).$$

Let $h_k = f - \Pi_k f$: since $|h_k| \leq \Delta_k f$, using (8.40) we get that

$$|S_n^{k-1}(h_k) - S_n^k(h_k)| \leq \sum_{i=1}^n |h_k(X_i^k) - h_k(X_i^{k-1})|$$

$$\leq \sum_{i=1}^n \mathrm{II}_{X_i^k \neq X_i^{k-1}} (\Delta_k f(X_i^k) + \Delta_k f(X_i^{k-1})). \qquad (8.51)$$

Hence, if n_k is the first integer multiple of $2q_k$ greater than n,

$$\mathbb{E}_{3,k} \leq n^{-1/2} \mathbb{E}\left(\sup_{g \in \mathcal{G}_k} \sum_{i=1}^{n_k} \mathrm{II}_{X_i^k \neq X_i^{k-1}} (g(X_i^k) + g(X_i^{k-1}))\right). \qquad (8.52)$$

In order to apply Lemma D.1, we now need to bound the Laplace transform of the random variables

$$T_{n_k,k}(g) = \sum_{i=1}^{n_k} \mathrm{II}_{X_i^k \neq X_i^{k-1}} (g(X_i^k) + g(X_i^{k-1})).$$

From Proposition 8.2, the random variable $T_{n_k,k}(g)$ is the sum of two random variables, which are sums of independent random variables with the same distribution as $T_{2q_k,k}(g)$ (with the exception of the last random variable). By Proposition 8.2(ii), the random variable $T_{2q_k,k}(g)$ has the same distribution as

$$T(g) = \sum_{i=1}^{q_k} \mathrm{II}_{X_i \neq X_i^*} (g(X_i) + g(X_i^*)).$$

Now, from (8.28), proceeding as in the proof of (8.30), we get that

$$\mathbb{E}(T(g)) = \sum_{i=1}^{q_k} \int_{\mathcal{X}} (b_i' + b_i^*) g \, dP = 2 \sum_{i=1}^{q_k} \int_{\mathcal{X}} b_i g \, dP.$$

Since $\|T(g)\|_\infty \leq 4q_k$, it follows that

$$2\mathbb{E}(T(g)) \leq \|g\|_{1,Q} \text{ and } \mathbb{E}(T^2(g)) \leq \|T(g)\|_\infty \mathbb{E}(T(g)) \leq 2q_k \|g\|_{1,Q}.$$

Hence, for any g in $\mathcal{G}_k$,

$$2\mathbb{E}(T(g)) \leq \delta_k, \quad \|T(g)\|_\infty \leq 2q_k \text{ and } \mathbb{E}(T^2(g)) \leq 2q_k \delta_k. \qquad (8.53)$$

Now, by (8.53) and Inequality (B.4),

$$\log \mathbb{E}(\exp(t T(g))) \leq (\delta_k t/2) + q_k \delta_k t^2/(1 - 8q_k t/3). \qquad (8.54)$$

Next, proceeding as in the proof (8.49) and applying (8.46), we get that

$$\mathbb{E}_{3,k} \leq n^{1/2}(2q_k)^{-1}\delta_k + \sqrt{8\delta_k H(\delta_k)} + 8n^{-1/2}q_k H(\delta_k) \leq 12\sqrt{\delta_k H(\delta_k)}. \qquad (8.55)$$

Finally,

$$\mathbb{E}_3 \leq \sum_{k=1}^{K} \mathbb{E}_{3,k} \leq 12 \sum_{k=1}^{K} \sqrt{\delta_k H(\delta_k)}. \qquad (8.56)$$

End of the proof of Theorem 8.3

For n large enough, $H(\delta) < n\delta$. Then $q_0 \geq 2$ and hence

$$q_0^2 H(\delta) \geq n\delta \text{ and } q_1^2 H(\delta) < n\delta,$$

whence $\delta \geq \delta_1$. Recall that $(\delta_k)_k$ satisfies the recursive equations

$$\delta_{k+1}^{-1} H(\delta_{k+1}) = 4\delta_k^{-1} H(\delta_k).$$

Let $G(x) = x^2 H(x)$. From the above equation,

$$\delta_k^3 G(\delta_{k+1}) = 4\delta_{k+1}^3 G(\delta_k).$$

Since G is nondecreasing, it follows that $\delta_k^3 \geq 4\delta_{k+1}^3$. Hence

$$2^{2/3}\delta_{k+1} \leq \delta_k \text{ for } k \geq 1 \text{ and } \delta_1 \leq \delta. \qquad (8.57)$$

Now, both Lemma 8.3 together with (8.42), (8.47), (8.50), (8.56) and (8.57) yield

$$\mathbb{E}(\sup_{f \in \mathcal{F}} |Z_n(f - \Pi_0 f)| \leq 4\sqrt{n}\beta_{q_0} + 6\sqrt{\delta_K H(\delta_K)} + 24 \sum_{k=1}^{K} \sqrt{\delta_k H(\delta_k)}.$$

Now, by (8.57) again, $\sqrt{\delta_k} \leq 3(\delta_k - \delta_{k+1})/\sqrt{\delta_k}$, whence

$$6\sqrt{\delta_K H(\delta_K)} + 24 \sum_{k=1}^{K} \sqrt{\delta_k H(\delta_k)} \leq$$

$$72\left(\sqrt{\delta_K H(\delta_K)} + \sum_{k=1}^{K-1}(\delta_k - \delta_{k+1})\sqrt{H(\delta_k)/\delta_k}\right). \qquad (8.58)$$

Since H is decreasing and $\delta_1 \leq \delta$, we thus get that

$$\mathbb{E}(\sup_{f \in \mathcal{F}} |Z_n(f - \Pi_0 f)| \le 4\sqrt{n}\beta_{q_0} + 72 \int_0^\delta (H(x)/x)^{1/2} dx. \qquad (8.59)$$

From the definition of q_0, $\sqrt{n}\beta_{q_0}$ converges to 0 as soon as $\lim_q q\beta_q = 0$. Hence, by (8.59), the stochastic equicontinuity holds true, which completes the proof of Theorem 8.3. ∎

Proof of Corollary 8.2. Let

$$C_\beta = 1 + 4 \sum_{i>0} \int_\mathcal{X} b_i dP.$$

Throughout this proof, $\mathbb{R}^d$ is equipped with the product order. As in the proof of Theorem 7.4, we may assume that the components of X_0 are uniformly distributed over $[0, 1]$. Let P denote the law of X_0. For t in $[0, 1]$, let

$$G_j(t) = Q(x = (x_1, \dots, x_d) \in \mathbb{R}^d : x_j \le t). \qquad (8.60)$$

For any (s, t) in $[0, 1]^d$ with $s < t$,

$$G_j(t) - G_j(s) = Q(x = (x_1, \dots, x_d) \in \mathbb{R}^d : s < x_j \le t).$$

Now Q is absolutely continuous with respect to P, which implies that the marginal distribution function G_j is continuous and has bounded variation. Furthermore, since $Q \ge P$, $G_j(t) - G_j(s) \ge t - s$. Hence G_j is a one to one continuous mapping from $[0, 1]$ onto $[0, C_\beta]$. Let

$$G(x_1, \dots x_d) = (C_\beta^{-1} G_1(x_1), \dots, C_\beta^{-1} G_d(x_d)) \text{ and } Y_i = G(X_i). \qquad (8.61)$$

The sequence of random variables $(Y_i)_i$ has the same β-mixing properties as the initial sequence $(X_i)_i$ Furthermore, from the definition of G, for any t in $[0, 1]^d$, $X_i \le t$ if and only if $G(X_i) \le G(t)$. Hence it is enough to prove that the empirical distribution function associated to $(Y_i)_i$ satisfies the uniform central limit theorem. Set

$$\Pi_K(t) = (2^{-K}[2^K t_1], \dots, 2^{-K}[2^K t_d]) \text{ and } \Pi_K^+(t) = (2^{-K}(1 + [2^K t_1]), \dots, 2^{-K}(1 + [2^K t_d])).$$

If $Y_0 = (Y_0^1, \dots, Y_0^d)$, then

$$\mathbb{P}(\Pi_K(t) \le Y_0 \le \Pi_K^+(t)) \le \sum_{j=1}^d \mathbb{P}([2^K t_j] \le 2^K Y_0^j \le [2^K t_j] + 1) \le dC_\beta 2^{-K}.$$

$$(8.62)$$

Hence the entropy with bracketing $H_{1,Q}$ associated to the new sequence $(Y_i)_i$ and the class of lower-left orthants satisfies $H_{1,Q}(x) = O(\log(1/x))$ as x tends to 0, which implies Corollary 8.2. ∎

Exercises

(1) Use Exercise 4 in Chap. 1 to prove that the map $f \to \|f\|_{2,\beta}$, which is defined in (8.21), is a norm on $L_{2,\beta}(P)$.

Problem In this problem, we will prove the uniform central limit theorem of Pollard (1982) for classes of functions satisfying a universal entropy condition, in a particular case.

Let $(X_i)_{i>0}$ be a sequence of independent random variables with values in some measure space $(\mathcal{X}, \mathcal{E})$, with common law P. Let $\mathcal{A}(\mathcal{X})$ be the set of probability measures on $\mathcal{X}$ with finite support. For any Q in $\mathcal{A}(\mathcal{X})$, we denote by d_Q the pseudodistance associated to the usual norm in $L^2(Q)$. Let $\mathcal{F}$ be a class of measurable functions from $\mathcal{X}$ into $[-1, 1]$. We set

$$H(x, \mathcal{F}) = \sup_{Q \in \mathcal{A}(\mathcal{X})} H(x, \mathcal{F}, d_Q), \tag{1}$$

where $H(x, \mathcal{F}, d_Q)$ is defined as in Definition 8.4. The function $x \to H(x, \mathcal{F})$ is called universal Koltchinskii–Pollard entropy. The universal entropy of $\mathcal{F}$ is said to be integrable if

$$\int_0^1 \sqrt{H(x, \mathcal{F})} \, dx < \infty. \tag{2}$$

The class $\mathcal{F}$ is said to fulfill the measurability condition (M) if there exists a countably generated and locally compact space $(K, \mathcal{B}(K))$ equipped with its Borel σ-field and a surjective map T from K onto $\mathcal{F}$ such that the map $(x, y) \to T(y)(x)$ is measurable with respect to the σ-fields $(\mathcal{X} \times K, \mathcal{E} \otimes \mathcal{B}(K))$ and $\mathcal{B}(\mathbb{R})$.

I. A symmetrization inequality.

Here $\mathcal{G}$ is a class of measurable functions from $\mathcal{X}$ into $[-1, 1]$, satisfying condition (M).

(1) Prove that the map

$$(x_1, \ldots, x_p, y_1, \ldots, y_q) \to \sup\{g(x_1) + \cdots + g(x_p) - g(y_1) - \cdots - g(y_q) : g \in \mathcal{G}\}$$

is universally measurable in the sense of Definition E.1, Annex E.

(2) Let $(X_i')_{i>0}$ be an independent copy of $(X_i)_{i>0}$. Let P_n be the empirical measure associated to $X_1, \ldots, X_n$, as defined in (1.37), and let P_n' denote the empirical measure associated to $X_1', \ldots, X_n'$. Prove that the variables in (3) are measurable and that

$$\mathbb{E}\left(\sup_{g \in \mathcal{G}} |P_n(g) - P(g)|\right) \leq \mathbb{E}\left(\sup_{g \in \mathcal{G}} |P_n(g) - P_n'(g)|\right). \tag{3}$$

Hint: apply Jensen's inequality conditionally to $X_1, \ldots, X_n$.

Let $(\varepsilon_i)_{i>0}$ be a sequence of symmetric independent signs, independent of the σ-field generated by $(X_i)_{i>0}$ and $(X'_i)_{i>0}$. Let $(X_i^s, X_i'^s)$ be defined by $(X_i^s, X_i'^s) = (X_i, X'_i)$ if $\varepsilon_i = 1$ and $(X_i^s, X_i'^s) = (X'_i, X_i)$ if $\varepsilon_i = -1$.

(3) Prove that the sequence $(X_i^s, X_i'^s)_i$ is a sequence of independent random variables with common law $P \otimes P$.

(4) Starting from (3), prove that

$$\mathbb{E}\left(\sup_{g \in \mathcal{G}} |P_n(g) - P'_n(g)|\right) = n^{-1}\mathbb{E}\left(\sup_{g \in \mathcal{G}} \left|\sum_{i=1}^{n} \varepsilon_i(g(X_i) - g(X'_i))|\right).$$

(5) Prove that

$$\mathbb{E}\left(\sup_{g \in \mathcal{G}} |P_n(g) - P(g)|\right) \leq 2n^{-1}\mathbb{E}\left(\sup_{g \in \mathcal{G}} \left|\sum_{i=1}^{n} \varepsilon_i g(X_i)|\right). \tag{4}$$

II. Stochastic equicontinuity of the symmetrized empirical process

Throughout Part II, we fix $(x_1, \ldots, x_n)$ in $\mathcal{X}^n$. We assume that the class of functions $\mathcal{G}$ satisfies the universal entropy condition (2). We set

$$\varphi(\sigma, \mathcal{G}) = \int_0^{\sigma} \sqrt{H(x, \mathcal{G})} \, dx. \tag{5}$$

Let $Q_n = n^{-1}(\delta_{x_1} + \cdots + \delta_{x_n})$ denote the empirical measure associated to $(x_1, \ldots, x_n)$. We define the empirical maximal variance V by $V = V(x_1, \ldots, x_n) = \sup\{Q_n(g^2) : g \in \mathcal{G}\}$. Let δ be any real in $]0, 1]$.

(1) Let I be a finite subset of $\mathcal{G}$ with cardinality $\exp(H) \geq 2$. Prove that

$$\mathbb{E}\left(\sup_{g \in I} \left|\sum_{i=1}^{n} \varepsilon_i g(x_i)|\right) \leq 2\sqrt{H} \sup_{g \in I}\left(\sum_{i=1}^{n} g^2(x_i)\right)^{1/2}. \tag{6}$$

(2) Prove that, for any nonnegative integer k, there exists a finite subset $\mathcal{G}_k$ of $\mathcal{G}$ with cardinality at most $\exp(H(2^{-k}\delta))$ and such that there exists some map Π_k from $\mathcal{G}$ into $\mathcal{G}_k$ satisfying the condition below:

$$d_{Q_n}(g, \Pi_k g) \leq 2^{-k}\delta \text{ for any } g \in \mathcal{G}.$$

(3) Prove that, for any function g in $\mathcal{G}$ and any integer l,

$$\left|\sum_{i=1}^{n} \varepsilon_i(g - \Pi_l g)(x_i)\right| \leq n2^{-l}\delta.$$

Infer from this inequality that

$$\mathbb{E}\left(\sup_{g\in\mathcal{G}}\left|\sum_{i=1}^{n}\varepsilon_i g(x_i)\right|\right) = \lim_{l\to\infty}\mathbb{E}\left(\sup_{g\in\mathcal{G}_l}\left|\sum_{i=1}^{n}\varepsilon_i g(x_i)\right|\right).$$

(4) Let $\delta_k = 2^{-k}\delta$. Prove that, for any g in $\mathcal{G}_l$, there exists a collection of functions $g_0,\ldots,g_l$ satisfying $g_l = g$ and $g_k = \Pi_k g_{k+1}$ for any integer k in $[0, l[$. Infer that

$$\mathbb{E}\left(\sup_{g\in\mathcal{G}_l}\left|\sum_{i=1}^{n}\varepsilon_i g(x_i)\right|\right) \leq 2\sum_{k=1}^{l}\delta_{k-1}\sqrt{nH(\delta_k)} + 2\sqrt{nH(\delta)V}.$$

(5) Prove that

$$\mathbb{E}\left(\sup_{g\in\mathcal{G}}\left|\sum_{i=1}^{n}\varepsilon_i g(x_i)\right|\right) \leq 8\sqrt{n}\varphi(\delta/2, \mathcal{G}) + 2\sqrt{nH(\delta)V}.$$

Infer from the above inequality that

$$\mathbb{E}\left(\sup_{g\in\mathcal{G}}\left|\sum_{i=1}^{n}\varepsilon_i g(X_i)\right|\right) \leq 8\sqrt{n}\varphi(\delta/2, \mathcal{G}) + 2\sqrt{nH(\delta)\mathbb{E}(V(X_1,\ldots,X_n))}. \quad (7)$$

III. Modulus of continuity of the normalized empirical process

Let H be a nonincreasing entropy function such that

$$\int_0^1 \sqrt{H(x)}dx < \infty \text{ and let } \varphi(\sigma) = \int_0^\sigma \sqrt{H(x)}dx.$$

Let $\mathcal{E}(\delta, P, H)$ be the set of classes of functions $\mathcal{G}$ from $\mathcal{X}$ into $[-1, 1]$, satisfying the measurability condition (M), such that $H(x, \mathcal{G}) \leq H(x)$ and $\sup\{P(g^2) : g \in \mathcal{G}\} \leq \delta^2$. We set

$$w(\delta) = \sup_{\mathcal{G}\in\mathcal{E}(\delta,P,H)} \mathbb{E}\left(\sup_{g\in\mathcal{G}}|Z_n(g)|\right).$$

(1) Let $\mathcal{G}$ be any class of functions in $\mathcal{E}(\delta, P, H)$. Prove that the class $\{g^2/2 : g \in \mathcal{G}\}$ still belongs to $\mathcal{E}(\delta, P, H)$. Infer that

$$w(\delta) \leq 16\varphi(\delta/2) + 4\sqrt{H(\delta)}\sqrt{\delta^2 + 2n^{-1/2}w(\delta)}. \quad (8)$$

Starting from (8), prove that

$$w(\delta) \leq 16\varphi(\delta) + 4w(\delta)\sqrt{H(\delta)/(n\delta^2)}.$$

(2) Prove that $w(\delta) \leq 32\varphi(\delta)$ for any positive δ satisfying $2^6 H(\delta) \leq n\delta^2$.

(3) Prove that the class

$$\mathcal{G}_\delta = \{(f - g)/2 : (f, g) \in \mathcal{F} \times \mathcal{F}, d_P(f, g) \leq \delta\}$$

belongs to $\mathcal{E}(\delta, P, H)$ for $H = H(., \mathcal{F})$. Infer that, if $n\delta^2 \geq 2^6 H(\delta)$, then

$$\mathbb{E}\left(\sup_{g \in \mathcal{G}_\delta} |Z_n(2g)|\right) \leq 64 \int_0^\delta \sqrt{H(x, \mathcal{F})}\, dx. \tag{9}$$

Apply Theorem 7.1 to conclude.

Chapter 9
Irreducible Markov Chains

9.1 Introduction

In this chapter, we are interested in the mixing properties of irreducible Markov chains with continuous state space. More precisely, our aim is to give conditions implying strong mixing in the sense of Rosenblatt (1956) or β-mixing. Here we mainly focus on Markov chains which fail to be ρ-mixing (we refer to Bradley (1986) for a precise definition of ρ-mixing). Let us mention that ρ-mixing essentially needs a spectral gap condition in L^2 for the transition probability kernel. This condition is often too restrictive for the applications in econometric theory or nonparametric statistics. However, these Markov models are irreducible. Thus, one can apply general results on irreducible Markov chains. We refer to Nummelin (1984) for more about irreducible Markov chains and to Meyn and Tweedie (1993) for a detailed exposition on Markov chains.

In Sect. 9.2, we give a brief exposition of the theory of irreducible Markov chains. In Sect. 9.3 we introduce regeneration techniques. Our exposition is based on the lecture notes of Nummelin (1984) and on Nummelin (1978). In Sect. 9.4, we give an example of an irreducible and positively recurrent Markov chain, and we apply the results of the previous sections to this example. For this example, we are able to give precise estimates of the strong mixing and the β-mixing coefficients. In Sect. 9.5, we give some relations between the integrability properties of return times, the rates of ergodicity, and the absolute regularity properties of the chain. Our exposition is based on papers by Lindvall (1979) and Tuominen and Tweedie (1994). In Sect. 9.6, we give relations between the rates of ergodicity, the absolute regularity coefficients and the strong mixing coefficients. Starting from papers of Bolthausen (1980, 1982b), we prove that, under some adequate assumptions, the coefficients of absolute regularity and the coefficients of strong mixing in the sense of Rosenblatt are of the same order of magnitude. Section 9.7 is devoted to the optimality of the central limit theorem of Chap. 4. The lower bounds are based on the example introduced in Sect. 9.4.

© Springer-Verlag GmbH Germany 2017
E. Rio, *Asymptotic Theory of Weakly Dependent Random Processes*,
Probability Theory and Stochastic Modelling 80,
DOI 10.1007/978-3-662-54323-8_9

9.2   Irreducible Markov Chains

In this section, we recall some classical results on irreducible Markov chains. We start
with the definition of the transition probability and the notion of irreducibility. Let
$(X, \mathcal{X})$ be a measurable space. Throughout this chapter, we assume that the σ-field
$\mathcal{X}$ is countably generated, which means that $\mathcal{X}$ is generated by a finite or a countable
family of sets. If X is topological, then $\mathcal{X}$ will be taken as the Borel σ-field of X, but
otherwise it may be arbitrary.

Definition 9.1 If $P : X \times \mathcal{X} \to \bar{\mathbb{R}}^+$ is such that

(i) for each A in $\mathcal{X}$, $P(\cdot, A)$ is a nonnegative measurable function on X,
(ii) for each x in X, $P(x, \cdot)$ is a nonnegative measure on $\mathcal{X}$,

then we call P a positive kernel. The kernel P is said to be finite (resp. σ-finite) if,
for any x in X, the measure $P(x, \cdot)$ is finite (resp. σ-finite). P is said to be bounded if
$\sup\{P(x, X) : x \in X\} < \infty$. P is said to be stochastic or to be a transition probability kernel if $P(x, X) = 1$ for any x in X. P is said to be substochastic if $P(x, X) \leq 1$
for any x in X.

The product $P_1 P_2$ of two positive kernels P_1 and P_2 is defined by

$$P_1 P_2(x, A) = \int_{\mathcal{X}} P_1(x, dy) P_2(y, A).$$

The powers P^n of P are defined by $P^0(x, A) = \delta_x(A) = \mathbb{1}_{x \in A}$ and $P^n = P P^{n-1}$. If
P is a transition probability kernel, then we call P^n the n-step transition probability
kernel.

Throughout, we call I the transition probability kernel defined by $I(x, A) =
\delta_x(A)$. If P is a transition probability kernel, then $G = \sum_{n \geq 0} P^n$ is called the potential
of P.

Let γ be a Radon measure and let f be a numerical measurable function. For any
x in X and any A in $\mathcal{X}$, let

$$\gamma P_1(A) = \int P_1(x, A)\gamma(dx), \; P_1 f(x) = \int f(y) P_1(x, dy) \text{ and } \gamma P_1(f) = \int P_1 f(x)\gamma(dx).$$

With these notations, for $f = \mathbb{1}_A$, $\gamma P_1(f) = \gamma P_1(A)$. We now define a relation on
$X \times \mathcal{X}$, called the communication structure.

Definition 9.2 Let (x, A) be an element of $X \times \mathcal{X}$. We say that A is accessible from
x under P if there exists a positive integer n such that $P^n(x, A) > 0$. In that case we
write $x \to A$. The set $\bar{A} = \{x \in X : L(x, A) > 0\}$ is the set of points from which A
is accessible.

Starting from Definition 9.2, we now give an extension of the notion of irreducibility to continuous state spaces.

Definition 9.3 Let φ be a positive and σ-finite measure on X such that $\varphi(X) > 0$. The stochastic kernel P is called φ-irreducible if $\bar{A} = X$ for any A in $\mathcal{X}$. The measures φ satisfying these conditions are called irreducibility measures under P. An irreducibility measure m under P is called a maximal irreducibility measure under P if any irreducibility measure φ under P is absolutely continuous with respect to m.

The proposition below, due to Tweedie (1974), gives a characterization of the maximal irreducibility measures.

Proposition 9.1 *Suppose that P is φ-irreducible under P. Then*

 (i) *There exists a maximal irreducibility measure m under P.*
 (ii) *Any irreducibility measure φ under P such that φP is absolutely continuous with respect to φ is maximal.*
(iii) *If $m(B) = 0$ then the set $B^+ = B \cup \{x \in X : x \to B\}$ satisfies $m(B^+) = 0$.*

Throughout the rest of this section, P is an irreducible stochastic kernel and m is a maximal irreducibility measure under P. The theorem below, due to Jain and Jamison (1967), gives a characterization of the irreducible stochastic kernels.

Theorem 9.1 *Let P be an irreducible stochastic kernel and let m be a maximal irreducibility measure under P. Then there exists a positive integer m_0, a measurable function s with values in $[0, 1]$ such that $m(s) > 0$ and a probability measure ν such that*

$$\mathcal{M}(m_0, s, \nu) \qquad P^{m_0}(x, A) \geq s(x)\nu(A) \ \text{for any} \ (x, A) \in X \times \mathcal{X}.$$

The substochastic kernel $(x, A) \to s(x)\nu(A)$ is denoted by $s \otimes \nu$.

Remark 9.1 (i) A positive measure φ is irreducible under P if and only if for any nonnegative function f such that $\varphi(f) > 0$, the potential G associated to P fulfills the following positivity condition: $PGf(x) > 0$ for any x in X.
(ii) In Theorem 9.1, one can assume that $\nu(s) > 0$ (see Sect. 2.3 in Nummelin (1984) for more about Theorem 9.1).

Assume now that $\mathcal{M}(m_0, s, \nu)$ is satisfied. From the above remark, P is ν-irreducible, since

$$PG \geq G(s \otimes \nu) = Gs \otimes \nu. \tag{9.1}$$

Consequently, if $\nu(f) > 0$ then $PGf(x) \geq Gs(x)\nu(f) > 0$ (the fact that $Gs(x) > 0$ is implied by the condition $m(s) > 0$). This fact together with Proposition 9.1(ii) lead to the remark below.

Remark 9.2 If $\mathcal{M}(1, s, \nu)$ is satisfied then $m = \sum_{n \geq 0} 2^{-1-n}\nu(P - s \otimes \nu)^n$ is a maximal irreducibility measure under P.

We now define the period of an irreducible Markov chain. Let (s, ν) satisfy condition $\mathcal{M}(m_0, s, \nu)$ for some positive integer m_0. Suppose furthermore that $\nu(s) > 0$. Let the set I be defined by

$$I = \{m \geq 1 : \mathcal{M}(m, \delta s, \nu) \text{ is fulfilled for some } \delta > 0\}.$$

The greatest common divisor of I is called the period of the chain. One can prove that the period does not depend on (s, ν). The chain is said to be aperiodic if $d = 1$. For example, if condition $\mathcal{M}(m_0, s, \nu)$ holds true with $m_0 = 1$, then the chain is aperiodic.

9.3   The Renewal Process of an Irreducible Chain

In this section, we consider a homogenous Markov chain with transition probability kernel $P(x, \cdot)$ and state space $X = [0, 1]$. Throughout the section, we assume that condition $\mathcal{M}(m_0, s, \nu)$ of Theorem 9.1 holds true for $m_0 = 1$, which ensures that the chain is irreducible and aperiodic. We will also assume that $\nu(s) > 0$.

Definition 9.4 Let the substochastic kernel Q be defined by $Q = P - s \otimes \nu$. The stochastic kernel Q_1 is defined from Q by

$$(1 - s(x))Q_1(x, A) = Q(x, A) \text{ if } s(x) < 1 \text{ and } Q(x, A) = \nu(A) \text{ if } s(x) = 1.$$

We now construct a stationary Markov chain with initial law μ and transition probability measure $P(x, \cdot)$ Let ζ_0 be a random variable with law μ. We assume that the underlying probability space is rich enough to contain a sequence $(U_i, \varepsilon_i)_{i \geq 0}$ of independent random variables with uniform law over $[0, 1]^2$, and that this random sequence is independent of ζ_0. For any x in $[0, 1]$ such that $s(x) < 1$, let F_x denote the distribution function of $Q_1(x, \cdot)$. Let F denote the distribution function of ν. The sequence $(\xi_n)_{n \geq 0}$ is defined by induction in the following way: $\xi_0 = \zeta_0$ and, for any nonnegative integer n,

$$\xi_{n+1} = \mathbb{1}_{s(\xi_n) \geq U_n} F^{-1}(\varepsilon_n) + \mathbb{1}_{s(\xi_n) < U_n} F_{\xi_n}^{-1}(\varepsilon_n). \tag{9.2}$$

By the Kolmogorov extension theorem, there exists a unique sequence $[\xi_n]_{n \geq 0}$ of random variables satisfying the above conditions. Furthermore, this sequence is a Markov chain. Now

$$\mathbb{P}(\xi_{n+1} \in A | \xi_n = x, U_n = u, \varepsilon_n = \varepsilon) = \mathbb{1}_{s(x) > u} \mathbb{1}_{F^{-1}(\varepsilon) \in A} + \mathbb{1}_{s(x) \leq u} \mathbb{1}_{F_x^{-1}(\varepsilon) \in A}.$$

Hence, integrating with respect to ε, we get that

$$\mathbb{P}(\xi_{n+1} \in A | \xi_n = x, U_n = u) = \mathbb{1}_{s(x) > u} \nu(A) + \mathbb{1}_{s(x) \leq u} Q_1(x, A).$$

Now, integrating on $[0, 1]$ with respect to u, we obtain that

$$\mathbb{P}(\xi_{n+1} \in A | \xi_n = x) = s(x)\nu(A) + (1 - s(x))Q_1(x, A) = P(x, A), \tag{9.3}$$

which proves that the transition probability kernel of this chain is $P(x, \cdot)$. Let us now define the renewal process associated to the so-constructed chain. The law of the renewal process will mainly depend on s and ν.

Definition 9.5 Let the sequence $(\eta_i)_{i \in \mathbb{N}}$ of random variables with values in $\{0, 1\}$ be defined by $\eta_i = \mathbb{1}_{U_i \leq s(\xi_i)}$. This sequence is called an incidence process associated to the chain $(\xi_i)_{i \in \mathbb{N}}$. The renewal times $(T_i)_{i \geq 0}$ are defined by

$$T_i = 1 + \inf\{n \geq 0 : \sum_{j=0}^{n} \eta_j = i + 1\}.$$

We set $\tau = T_0$ and $\tau_i = T_{i+1} - T_i$ for $i \geq 0$.

Let $\mathbb{P}_\mu$ be the law of the chain with transition probability kernel P and initial law μ. When $\mu = \delta_x$, we denote this law by P_x. By definition of τ,

$$\lambda Q^n(s) = \mathbb{P}_\lambda(\tau = n + 1) \quad \text{and} \quad \lambda Q^n(1) = \mathbb{P}_\lambda(\tau > n). \tag{9.4}$$

Let us make some comments about (9.4). For any initial law λ,

$$\lambda P^n - \lambda Q^n = \sum_{k=1}^{n} \lambda Q^{n-k}(P - Q) P^{k-1}.$$

Since $P - Q = s \otimes \nu$, $\lambda Q^{n-k}(P - Q) P^{k-1} = \lambda Q^{n-k}(s) \nu P^{k-1}$, which leads to the identity

$$\lambda P^n = \sum_{k=1}^{n} \lambda Q^{n-k}(s) \nu P^{k-1} + \lambda Q^n. \tag{9.5}$$

The equality of total masses in (9.5) yields

$$\sum_{k=1}^{n} \lambda Q^{n-k}(s) + \lambda Q^n(1) = 1. \tag{9.6}$$

The last equality corresponds to the trivial identity

$$\mathbb{P}_\lambda(\tau > n) + \sum_{k=1}^{n} \mathbb{P}_\lambda(\tau = k) = 1.$$

The identity (9.5) provides more information, and will be used again in the following sections.

We now give the definition of recurrence and classical results on the recurrence properties of irreducible chains.

Definition 9.6 Let $(\xi_i)_{i\geq 0}$ be an irreducible Markov chain with maximal irreducibility measure m. The chain is said to be recurrent if, for any B in $\mathcal{X}^+ = \{A \in \mathcal{X} : m(A) > 0\}$,

$$h_B^\infty(x) = \mathbb{P}_x\left(\sum_{k\geq 0} \mathbb{1}_{\xi_k \in B} = \infty\right) = 1 \ m\text{-almost everywhere.}$$

One can prove that the chain with transition probability P starting from $\xi_0 = x$ is recurrent if and only if T_i is finite almost surely for any nonnegative integer i. Consequently, applying (9.4), we get the lemma below.

Lemma 9.1 *The irreducible Markov chain $(\xi_i)_{i\geq 0}$ is recurrent if and only if*

$$\lim_{n\to\infty} \nu Q^n(1) = 0 \ \text{and} \ \lim_{n\to\infty} \delta_x Q^n(1) = 0 \ m\text{-almost everywhere.}$$

We now prove that the second condition appearing in Lemma 9.1 can be removed. Assume that $\lim_n \nu Q^n(1) = 0$. By (9.6) applied to $\lambda = \delta_x$,

$$\sum_{l=0}^{n-1} Q^l s(x) + \delta_x Q^n(1) = 1.$$

Let the nonnegative kernel G_Q be defined by $G_Q = \sum_{n\geq 0} Q^n$. From the above identity, $G_Q s(x) \leq 1$ and $G_Q s(x) = 1$ if and only if $\lim_n \delta_x Q^n(1) = 0$, which is equivalent to $\mathbb{P}_x(\tau = \infty) = 0$. Now, since $\lim_n \nu Q^n(1) = 0$, $\mathbb{P}_\nu(\tau = \infty) = 0$. Consequently, if $G_Q s(x) = 1$, then the chain with transition P starting from x is recurrent ($h_B(x) = 1$ for any B in $\mathcal{X}^+$). Let m be the maximal irreducibility measure defined in Remark 9.2. The Markov chain with transition P is recurrent if and only if $G_Q s(x) = 1$ m-almost everywhere. Since $G_Q s(x) \leq 1$, this equality holds m-almost everywhere if and only if

$$m G_Q(s) = \int_X G_Q s(x) m(dx) = \int_X 1.m(dx) = m(1).$$

Consequently, the chain is recurrent if and only if

$$\sum_{n\geq 0} 2^{-1-n} \sum_{p\geq n} \nu Q^p(s) = \sum_{n\geq 0} 2^{-1-n} \nu Q^n(1). \tag{9.7}$$

Now

$$\nu Q^n(1) - \sum_{p\geq n} \nu Q^p(s) = \mathbb{P}_\nu(\tau = \infty). \tag{9.8}$$

Hence the equality (9.7) holds if and only if $\mathbb{P}_\nu(\tau = \infty) = \lim_n \nu Q^n(1) = 0$. Thus we have proved the proposition below.

Proposition 9.2 *Let m be the maximal irreducibility measure of Remark 9.2. The irreducible chain $(\xi_i)_{i\geq 0}$ is recurrent if and only if*

$$\lim_{n\to\infty} \nu Q^n(1) = 0.$$

9.4 Mixing Properties of Positively Recurrent Markov Chains: An Example

Throughout this section $X = [0, 1]$. For Markov chains, the strong mixing coefficients defined in (2.1) and the β-mixing coefficients defined in Definition 8.3 satisfy

$$\alpha_n = \sup_{k\in T} \alpha(\sigma(X_k), \sigma(X_{k+n})) \quad \text{and} \quad \beta_n = \sup_{k\in T} \beta(\sigma(X_k), \sigma(X_{k+n})). \tag{9.9}$$

We refer to Davydov (1973) and to Bradley (1986) for a proof of this result.

Let us consider an irreducible Markov chain. Suppose there exists a pair (s, ν) satisfying condition $\mathcal{M}(m_0, s, \nu)$ with $m_0 = 1$. Then the chain is aperiodic. Let $Q = P - s \otimes \nu$. Assume furthermore that the positive measure $\sum_{n\geq 0} \nu Q^n$, which is usually called the Pitman occupation measure (see Pitman (1974) for more about this measure), has a finite total mass. Then the probability measure

$$\pi = \left(\sum_{n\geq 0} \nu Q^n(1)\right)^{-1} \sum_{n\geq 0} \nu Q^n \tag{9.10}$$

is an invariant law under P. Furthermore, the chain is recurrent, the renewal times $(\tau_i)_{i\geq 0}$ are integrable and the return times in a recurrent set A (A is recurrent if $m(A) > 0$) are also integrable. In that case the chain is said to be positively recurrent.

In this section, we will introduce an additional assumption which provides nice estimates of the mixing coefficients. This assumption will be called the excessivity assumption. Under this assumption the lemma below provides a rate of convergence to the invariant law π.

Lemma 9.2 *Let P be an irreducible transition probability kernel satisfying $\mathcal{M}(1, s, \nu)$ and λ an initial probability law. Suppose that the following assumption holds true:*

$$\lambda P^l(s) \geq \pi(s) \text{ for any } l \geq 0. \qquad \mathcal{H}(\lambda, s)$$

Then, for any positive integer n,

$$\|\lambda P^n - \pi\| \leq 2\pi Q^n(1).$$

Proof Using the decomposition

$$\lambda P^n = \sum_{k=1}^{n} \lambda P^{n-k}(P - Q)Q^{k-1} + \lambda Q^n$$

and proceeding as in the proof of (9.5), we get that

$$\lambda P^n = \sum_{k=1}^{n} \lambda P^{k-1}(s)\nu Q^{n-k} + \lambda Q^n. \tag{9.11}$$

Apply now (9.11) with $\lambda = \pi$: since π is an invariant measure, we get that

$$\pi = \sum_{k=1}^{n} \pi(s)\nu Q^{n-k} + \pi Q^n.$$

Hence

$$\lambda P^n - \pi = \sum_{k=1}^{n} (\lambda P^{k-1}(s) - \pi(s))\nu Q^{n-k} + \lambda Q^n - \pi Q^n. \tag{9.12}$$

Since $\lambda P^{k-1}(s) - \pi(s) \geq 0$, the measure

$$\mu = \sum_{k=1}^{n} (\lambda P^{k-1}(s) - \pi(s))\nu Q^{n-k} + \lambda Q^n$$

is a nonnegative measure. Now $\lambda P(X) = \pi(X) = 1$, which implies that $\mu(X) = \pi Q^n(X) = \pi Q^n(1)$. Hence the decomposition (9.12) ensures that $\lambda P^n - \pi$ is the difference of two nonnegative measures with masses $\pi Q^n(1)$, which completes the proof of Lemma 9.2. ∎

Starting from Lemma 9.2, we now bound the β-mixing coefficients of the stationary chain with transition P.

Proposition 9.3 *Let P be a transition probability kernel satisfying the assumptions of Lemma 9.2, with $\lambda = \nu$. Then, for any positive integer n,*

$$\beta_n = \int_{\mathcal{X}} \|\delta_x P^n - \pi\| \pi(dx) \leq 2\pi Q^n(1) + 2\sum_{k=1}^{n} \pi Q^{k-1}(s)\pi Q^{n-k}(1).$$

Remark 9.3 Let $(\xi_i)_{i \geq 0}$ be the stationary chain with transition P and τ be the first renewal time, as defined in Definition 9.5. Let τ' be an independent copy of τ. Then Proposition 9.3 is equivalent to the upper bound $\beta_n \leq \mathbb{P}(\tau + \tau' > n)$.

Proof of Proposition 9.3 Applying (9.5) with $\lambda = \delta_x$, we get that

$$\delta_x P^n = \sum_{k=1}^{n} Q^{k-1}s(x)\nu P^{n-k} + \delta_x Q^n. \tag{9.13}$$

Next the equality of masses in (9.13) yields

$$\sum_{k=1}^{n} Q^{k-1} s(x) + \delta_x Q^n(1) = 1.$$

Consequently,

$$\delta_x P^n - \pi = \sum_{k=1}^{n} Q^{k-1} s(x)(\nu P^{n-k} - \pi) + \delta_x Q^n - \delta_x Q^n(1)\pi,$$

which ensures that

$$\|\delta_x P^n - \pi\| = \sum_{k=1}^{n} Q^{k-1} s(x) \|\nu P^{n-k} - \pi\| + 2\delta_x Q^n(1). \tag{9.14}$$

Integrating (9.14) with respect to π, we get that

$$\int_{\mathcal{X}} \|\delta_x P^n - \pi\| \pi(dx) \leq \sum_{k=1}^{n} \pi Q^{k-1}(s) \|\nu Q^{n-k} - \pi\| + 2\pi Q^n(1).$$

Applying Lemma 9.2 with $\lambda = \nu$ we then obtain Proposition 9.3. ∎

To conclude this section, we give an example of a kernel satisfying the assumptions of Proposition 9.3.

Lemma 9.3 *Let ν be an atomless law and s be a measurable function with values in $]0, 1]$ such that $\nu(s) > 0$. Suppose furthermore that*

$$\int_{\mathcal{X}} \frac{1}{s(x)} \nu(dx) < \infty. \tag{a}$$

Let $P(x, \cdot) = s(x)\nu + (1 - s(x))\delta_x$. Then P is aperiodic, positively recurrent and satisfies $\mathcal{H}(\nu, s)$.

Remark 9.4 Since ν is an atomless law, the renewal times are observable.

Proof Clearly $Q = P - s \otimes \nu = (1 - s(x))\delta_x$, whence $\nu Q^n = (1 - s(x))^n \nu$. It follows that the Pitman occupation measure is equal to $s^{-1}\nu$. By assumption (a), this measure has a finite total mass, which ensures that the chain is positively recurrent. Furthermore,

$$\pi = \left(\int_{\mathcal{X}} \frac{1}{s(x)} \nu(dx) \right)^{-1} \frac{1}{s(x)} \nu \tag{9.15}$$

is the unique invariant law under P.

Let $a_0 = 1$ and $a_k = \nu P^{k-1}(s) - \pi(s)$. The equality of masses in (9.12) yields

$$a_n + \sum_{k=0}^{n-1} a_k \nu Q^{n-k}(1) = \pi Q^n(1). \tag{9.16}$$

Set $t(x) = 1 - s(x)$. From the convexity $l \to \log \mathbb{E}_\nu(t^l)$,

$$\nu Q^{n-k}(1) = \mathbb{E}_\nu(t^{n-k}) \le \mathbb{E}_\nu(t^{n-k-1}) \frac{\mathbb{E}_\nu(t^n)}{\mathbb{E}_\nu(t^{n-1})}. \tag{9.17}$$

Hence,

$$\pi Q^n(1) \le a_n + \frac{\mathbb{E}_\nu(t^n)}{\mathbb{E}_\nu(t^{n-1})} \sum_{k=0}^{n-1} a_k \nu Q^{n-k-1}(1) = a_n + \frac{\mathbb{E}_\nu(t^n)\pi Q^{n-1}(1)}{\mathbb{E}_\nu(t^{n-1})}, \tag{9.18}$$

which ensures that

$$a_n \mathbb{E}_\nu(t^{n-1}) \ge \pi Q^n(1)\mathbb{E}_\nu(t^{n-1}) - \mathbb{E}_\nu(t^n)\pi Q^{n-1}(1). \tag{9.19}$$

It follows that $\mathcal{H}(\nu, s)$ is implied by the weaker condition

$$\mathbb{E}_\pi(t^n)\mathbb{E}_\nu(t^{n-1}) - \mathbb{E}_\nu(t^n)\mathbb{E}_\pi(t^{n-1}) \ge 0.$$

From (9.15) and the fact that $1/s = \sum_{l \ge 0} t^l$, the last condition holds if and only if

$$\sum_{k \ge n} \left(\mathbb{E}_\nu(t^k)\mathbb{E}_\nu(t^{n-1}) - \mathbb{E}_\nu(t^n)\mathbb{E}_\nu(t^{k-1}) \right) \ge 0.$$

Now, from the convexity of $l \to \log \mathbb{E}_\nu(t^l)$,

$$\mathbb{E}_\nu(t^k)\mathbb{E}_\nu(t^{n-1}) - \mathbb{E}_\nu(t^n)\mathbb{E}_\nu(t^{k-1}) \ge 0 \text{ for any } k \ge n,$$

which ensures that each term in the above sum is nonnegative. Hence (9.19) holds true, which completes the proof of Lemma 9.3. ∎

Starting from Proposition 9.3, we now give estimates of the β-mixing coefficients for the transition P defined in Lemma 9.3 in the stationary case.

Proposition 9.4 *Let ν be an atomless law on $]0, 1]$ and s be a function with values in $]0, 1]$ such that $\nu(s) > 0$. Let $P(x, \cdot) = s(x)\nu + (1 - s(x))\delta_x$. Assume that the assumption (a) of Lemma 9.3 holds. Let $(\xi_i)_{i \ge 0}$ be the stationary chain with transition probability kernel P. Then the stationary law is the unique invariant law π defined by (9.16). Now, let $\tau = \inf\{i > 0 : \xi_i \ne \xi_{i-1}\}$ and τ' be an independent copy of τ. Then, for any positive n,*

$$\mathbb{P}(\tau > n) \le \beta_n \le \mathbb{P}(\tau + \tau' > n).$$

Proof The upper bound comes from Proposition 9.3 together with Remark 9.3. We now prove the lower bound. From (9.12) applied with $\lambda = \delta_x$,

$$\delta_x P^n - \pi = \sum_{k=1}^{n} (P^{k-1} s(x) - \pi(s)) \nu Q^{n-k} + (1 - s(x))^n \delta_x - \pi Q^n.$$

Since the measures νQ^{n-k} and πQ^n are atomless, it follows that

$$\|\delta_x P^n - \pi\| \geq (1 - s(x))^n.$$

Integrating this lower bound with respect to the invariant law π then yields the desired result. ∎

In Sect. 9.7, we will apply Proposition 9.4 to prove the optimality of the strong mixing condition of Theorem 4.2. In the forthcoming sections, we give links between ergodicity, regularity and strong mixing.

9.5 Small Sets, Absolute Regularity and Strong Mixing

In this section, we give relations between the return times in small sets in the sense of Nummelin and the various mixing coefficients.

Definition 9.7 Let P be an irreducible and recurrent transition probability kernel, m be a maximal irreducibility measure and D be a measurable set such that $m(D) > 0$. A set D is called a small set if there exists a positive integer m, a positive constant ρ and a probability measure ν such that $P^m(x, .) \geq \rho \mathbb{I}_D(x) \nu$. The chain is said to be Doeblin recurrent if X is a small set. Then the above condition is called Doeblin's condition.

The small sets are called *C-sets* by Orey (1971) and *small sets* by Nummelin (1984). They differ from the *petite sets* defined in Meyn and Tweedie (1993).

We now prove that the Doeblin recurrent chains are geometrically uniformly mixing. This result is essentially due to Doeblin (1938). Here we give a proposition which can be found in Ueno (1960).

Proposition 9.5 *Let P be a probability transition kernel satisfying Doeblin's condition with $m = N$. Then, for any measurable set A such that $\nu(A) > 0$, any (x, x') in $X \times X$ and any positive integer k,*

$$|P^{Nk}(x, A) - P^{Nk}(x', A)| \leq (1 - \rho)^k. \tag{i}$$

Furthermore, there exists a unique invariant probability law π under P. The chain $(\xi_i)_{i \in \mathbb{Z}}$ with probability transition kernel P and initial law π satisfies

$$\varphi_{Nk} \leq (1 - \rho)^k. \tag{ii}$$

Proof We prove (i) by induction on k. For $k = 1$,

$$P^N(x, A) - P^N(x', A) = (P^N(x, A) - \rho\nu(A)) - (P^N(x', A) - \rho\nu(A)).$$

Hence

$$|P^N(x, A) - P^N(x', A)| \le 1 - \rho.$$

Suppose that (i) holds true at range k. Write

$$P^{Nk+N}(x, A) - P^{Nk+N}(x', A) = \int_X (P^N(y, A) - \rho\nu(A))(P^{Nk}(x, dy) - P^{Nk}(x', dy)).$$

Now the function $y \to P^N(y, A) - \rho\nu(A)$ takes its values in $[0, 1 - \rho]$. Let

$$B_u = \{y \in X : P^N(y, A) - \rho\nu(A) > u\}.$$

Then

$$P^N(y, A) - \rho\nu(A) = \int_0^{1-\rho} \amalg_{B_u}(y)du.$$

Therefore, by Fubini's theorem,

$$P^{Nk+N}(x, A) - P^{Nk+N}(x', A) = \int_0^{1-\rho}(P^{Nk}(x, B_u) - P^{Nk}(x', B_u))du.$$

Now $|P^{Nk}(x, B_u) - P^{Nk}(x', B_u)| \le (1 - \rho)^k$ under the induction hypothesis. Consequently, if (i) holds true at range k, then (i) holds true at range $k + 1$. Thus, by induction on k, (i) holds true for any positive integer k.

We now prove (ii). We start by noting that

$$\pi_0 = \nu + \nu(P^N - \rho\nu) + \cdots + \nu(P^N - \rho\nu)^k + \cdots$$

is invariant under P^N, since $\pi_0(P^N - \rho\nu) = \pi_0 - \nu$ (the total mass of π_0 is equal to $1/\rho$). Next the measure $\pi_1 = \pi_0 + \pi_0 P + \cdots + \pi_0 P^{N-1}$ is invariant under P. We then set $\pi = \pi_1/\pi_1(\mathcal{X})$. π is an invariant law under P. Now, by (i), for any measurable set A,

$$P^{Nk}(x', A) - \pi P^{Nk}(A) = \int_{\mathcal{X}} (P^{Nk}(x', A) - P^{Nk}(x, A))\pi(dx) \le (1 - \rho)^k,$$

which ensures that $\varphi_{Nk} \le (1 - \rho)^k$.

We now prove that π is unique. If π' is an invariant law, then

$$\pi'(A) - \pi(A) = \pi' P^{Nk}(A) - \pi P^{Nk}(A)$$
$$= \iint (P^{Nk}(x, A) - P^{Nk}(x', A))\pi \otimes \pi'(dx, dx').$$

Hence $|\pi'(A) - \pi(A)| \leq (1 - \rho)^k$ for any natural integer k, which implies that $\pi = \pi'$. ∎

Suppose now that the chain fails to be Doeblin recurrent. Then the rate of convergence to the invariant measure depends on the initial law and the chain fails to be uniformly mixing. Throughout the rest of this section, we are interested in the relations between the integrability properties of the renewal times and the rates of mixing for non-uniformly mixing Markov chains. Our aim is to remove the excessivity assumption of Sect. 9.4. We will extend results of Bolthausen (1980, 1982b) to general rates of mixing. Our extensions are based on a proposition of Lindvall (1979) which gives a link between coupling and regeneration times. Thus, we start by introducing the coupling method, which goes back to Doeblin (1938). Our exposition comes from Pitman's (1974) paper.

Let us consider two initial laws μ and λ. We define the chain $(\xi_i, \xi_i')_{i \geq 0}$ on $X \times X$ as follows: the initial law of the chain is $\mu \otimes \lambda$, and the transition probability kernel is $P \otimes P$. Then $(\xi_i)_{i \geq 0}$ is a Markov chain with initial law μ and transition probability kernel P and $(\xi_i')_{i \geq 0}$ is a Markov chain with initial law λ and transition P. Furthermore, from the definition, these chains are independent. Let $(\eta_i)_{i \geq 0}$ and $(\eta_i')_{i \geq 0}$ denote the incidence processes associated to the chains $(\xi_i)_{i \geq 0}$ and $(\xi_i')_{i \geq 0}$ (see Definition 9.5 for the definition). Set

$$T = 1 + \inf\{i \geq 0 : \eta_i = \eta_i' = 1\}. \tag{9.20}$$

Then ξ_T and ξ_T' have the distribution ν. Furthermore, ξ_T is independent of $(\xi_i)_{i < T}$ and ξ_T' is independent of $(\xi_i')_{i < T}$. We now define the coupled chain $(\xi_i'')_{i \geq 0}$ by $\xi_i'' = \xi_i'$ for $i < T$ and $\xi_i'' = \xi_i$ for $i \geq T$. By the Markov property, this new chain is a Markov chain with initial law λ. Furthermore, $\xi_i'' = \xi_i$ for $i \geq T$. We call T the coupling time of the chains. From the definition of the coupling time,

$$\int_{X \times X} \|\delta_x P^n - \delta_y P^n\| \, \mu \otimes \lambda(dx, dy) \leq 2\mathbb{P}_{\mu \otimes \lambda}(T > n). \tag{9.21}$$

We refer to Pitman (1974) for a proof of (9.21). If the kernel P is positively recurrent and if π is the invariant law, then, applying (9.21) with $\lambda = \mu = \pi$, we get that

$$\beta_n \leq \mathbb{P}_{\pi \otimes \pi}(T > n). \tag{9.22}$$

Consequently, the rate of β-mixing is closely related to the tail of the coupling time. In order to give more precise quantitative results, let us now introduce some classes of increasing functions.

Definition 9.8 Let Λ_0 be the class of nondecreasing functions ψ from $\mathbb{N}$ into $[2, +\infty[$ such that the sequence $((\log \psi(n))/n)_n$ is nonincreasing and converges to 0. For ψ in Λ_0, define the cumulative function ψ_0 associated to ψ by $\psi^0(k) = \sum_{i=0}^{k-1} \psi(i)$.

The proposition below, due to Lindvall (1979), generalizes a previous result of Pitman (1974). We refer to Lindvall (1979) for a proof of this result.

Proposition 9.6 *Let P be a stochastic kernel. Assume that P is irreducible, aperiodic, positively recurrent, and satisfies condition $\mathcal{M}(1, s, \nu)$. Let ψ be an element of Λ_0. If*

$$\mathbb{E}_\mu(\psi(T_0)) < \infty, \quad \mathbb{E}_\lambda(\psi(T_0)) < \infty \text{ and } \mathbb{E}_\nu(\psi^0(T_0)) < \infty, \tag{a}$$

then $\mathbb{E}_{\mu\otimes\lambda}(\psi(T)) < \infty$.

In the stationary case $\lambda = \mu = \pi$, by (9.4) together with the definition of π,

$$\mathbb{E}_\nu(T_0)\mathbb{P}_\pi(T_0 = n + 1) = \mathbb{P}_\nu(T_0 > n). \tag{9.23}$$

Hence,

$$\mathbb{E}_\nu(\psi^0(T_0)) = \mathbb{E}_\nu(T_0) \sum_{k \geq 0} \mathbb{P}_\pi(T_0 = k + 1)\psi(k) \leq \mathbb{E}_\pi(\psi(T_0)). \tag{9.24}$$

From (9.22) and Proposition 9.6, we now get the corollary below.

Corollary 9.1 *Let P be a stochastic kernel. Assume that P is irreducible, aperiodic, positively recurrent, and satisfies condition $\mathcal{M}(1, s, \nu)$. Let π denote the invariant probability law. Then, for any ψ in Λ_0 such that $\mathbb{E}_\pi(\psi(T_0)) < \infty$,*

$$\int_0^1 \psi(\beta^{-1}(u))du < +\infty.$$

Remark 9.5 From Corollary 9.1, one can derive the following result. Suppose that T_0 has a finite Laplace transform in a neighborhood of 0. Let U be a random variable with uniform law over $[0, 1]$. Then $\beta^{-1}(U)$ has a finite Laplace transform in another neighborhood of 0. To prove this fact, suppose that $\mathbb{E}(\exp(\varepsilon\beta^{-1}(U))) = \infty$ for any positive ε. Then one can construct a function ψ in Λ_0 such that $\mathbb{E}(\psi(\beta^{-1}(U))) = \infty$. For this function $\mathbb{E}_\pi(\psi(T_0)) = \infty$. Now, since T_0 has a finite Laplace transform in a neighborhood of 0, $\mathbb{E}_\pi(\psi(T_0)) < \infty$ for any ψ in Λ_0, which leads to a contradiction. Hence geometric ergodicity implies geometric β-mixing.

We now give quantitative relations concerning return times in small sets and absolute regularity coefficients. Our results are derived from the paper of Tuominen and Tweedie (1994).

Definition 9.9 Let f be a measurable function from X into $[1, \infty]$ and let m be a signed measure. The f-variation of m is defined by $\|m\|_f = \sup\{|m(g)| : |g| \leq f\}$.

Definition 9.10 Let $\tau_D = \inf\{n > 0 : \xi_n \in D\}$. The aperiodic and irreducible chain $(\xi_i)_{i \geq 0}$ is said to be (f, ψ)-ergodic if and only if there exists a small set D such that

$$\sup_{x \in D} \mathbb{E}_x \left(\sum_{i=o}^{\tau_D - 1} \psi(i) f(\xi_i) \right) < \infty. \tag{9.25}$$

We now give the ergodicity criterion of Tuominen and Tweedie (1994).

Theorem 9.2 *Let P be a stochastic kernel. Suppose that P is irreducible and aperiodic. Let ψ be an element of Λ_0. The chain $(\xi_i)_{i \geq 0}$ with kernel P is (f, ψ)-ergodic if and only if there exists a sequence $(V_n)_{n \geq 0}$ of measurable functions from X into $\bar{\mathbb{R}}^+$, a small set C and a positive constant b such that V_0 is bounded over C, $V_0(x) = \infty$ implies $V_1(x) = \infty$, and, for any $n \geq 0$,*

$$\psi(n) f \leq V_n - P V_{n+1} + b \psi(n) \mathbb{1}_C. \tag{9.26}$$

Proof Here we prove that (9.26) implies the (f, ψ)-ergodicity. We refer to Tuominen and Tweedie (1994) for a complete proof and for more details. Applying (9.26) with $n = i$ to $x = \xi_i$, summing on i from $i = 0$ to $i = \tau_C - 1$ we get that

$$\mathbb{E}_x \left(\sum_{i=o}^{\tau_C - 1} \psi(i) f(\xi_i) \right) \leq \sum_{i=o}^{\tau_C - 1} \mathbb{E}_x (V_i(\xi_i) - P V_{i+1}(\xi_i)) + b \psi(0) \mathbb{1}_{x \in C}.$$

Now $\mathbb{E}_x(P V_{i+1}(\xi_i)) = \mathbb{E}_x(V_{i+1}(\xi_{i+1}))$. Hence the above inequality ensures that

$$\mathbb{E}_x \left(\sum_{i=o}^{\tau_C - 1} \psi(i) f(\xi_i) \right) \leq V_0(x) + b \psi(0) \mathbb{1}_{x \in C}. \tag{9.27}$$

Since V_0 is uniformly bounded over C, (9.27) implies (9.25). ∎

We now give applications of the (f, ψ)-ergodicity to estimates of the absolute regularity coefficients of the chain. We refer to Theorems 3.6(i) and 4.3 in Tuominen and Tweedie (1994) for more about this subject.

Theorem 9.3 *Let P be a stochastic kernel. Suppose that P is irreducible and aperiodic. Let ψ be an element of Λ_0. Assume that the chain $(\xi_i)_{i \geq 0}$ with kernel P is (f, ψ)-ergodic. Then the chain is positively recurrent and, if π denotes the unique stationary law, then*

$$\sum_{n=0}^{\infty} \psi(n) \int_X \| P^n(x, .) - \pi \|_f \pi(dx).$$

In particular, if $f = 1$, then $\mathbb{E}(\psi^0(\beta^{-1}(U))) < \infty$ (here U has the uniform law over $[0, 1]$).

9.6   Rates of Strong Mixing and Rates of Ergodicity
of Irreducible Chains

In this section, we give relations between the strong mixing coefficients and the integrability properties of renewal times. We will prove that the tails of the random variables $\alpha^{-1}(U)$, T_0 and $\beta^{-1}(U)$ have the same order of magnitude, which implies, in particular, that the strong mixing and the β-mixing coefficients are of the same order for irreducible, aperiodic and positively recurrent Markov chains.

We start by giving some relations between the strong mixing coefficients of the chain $(\xi_i)_{i \geq 0}$ equipped with the usual filtration $(\mathcal{F}_k)_k$ defined by $\mathcal{F}_k = \sigma(\xi_i : i \leq k)$ and the strong mixing coefficients of the extended chain $(\xi_i, U_i)_{i \geq 0}$ defined in Sect. 9.3 by Eq. (9.2). Our lemma is inspired by Lemma 5 in Bolthausen (1982b).

Lemma 9.4 *Let $(\tilde{\alpha}_n)_{n \geq 0}$ and $(\tilde{\beta}_n)_{n \geq 0}$ denote respectively the sequences of strong mixing and β-mixing coefficients of the completed chain $(\xi_i, U_i)_{i \geq 0}$. Then, for any positive integer n,*

$$\tilde{\alpha}_{n+1} \leq \alpha_n \leq \tilde{\alpha}_n \quad and \quad \tilde{\beta}_{n+1} \leq \beta_n \leq \tilde{\beta}_n.$$

Proof If $\mathcal{C}$ is a σ-field independent of $\mathcal{A} \vee \mathcal{B}$, then, as proved in Bradley (1986),

$$\alpha(\mathcal{A}, \mathcal{B} \vee \mathcal{C}) = \alpha(\mathcal{A}, \mathcal{B}).$$

Now U_{k+n} is independent of $(\xi_{k+n}, \xi_{k-1}, U_{k-1})$. Hence, by (1.10b),

$$\tilde{\alpha}_{n+1} = \sup_{k > 0} \ \sup_{B \in \mathcal{B}(\mathbb{R})} \ \mathbb{E}|\mathbb{P}(\xi_{k+n} \in B | \mathcal{F}_{k-1}) - \mathbb{P}(\xi_{k+n} \in B)|.$$

Now

$$\mathbb{E}|\mathbb{P}(\xi_{k+n} \in B | \mathcal{F}_{k-1}) - \mathbb{P}(\xi_{k+n} \in B)| \leq \mathbb{E}|\mathbb{P}(\xi_{k+n} \in B | \mathcal{F}_{k-1} \vee \sigma(\xi_k)) - \mathbb{P}(\xi_{k+n} \in B)|.$$

Since ξ_{k+n} is a measurable deterministic function of ξ_k and $(U_i, \varepsilon_i)_{i \in [k, k+n[}$,

$$\mathbb{P}(\xi_{k+n} \in B | \mathcal{F}_{k-1} \vee \sigma(\xi_k)) = \mathbb{P}(\xi_{k+n} \in B | \sigma(\xi_k)).$$

It follows that $\tilde{\alpha}_{n+1} \leq \alpha_n$. The proof of the inequality $\beta_n \leq \tilde{\beta}_{n+1}$ is similar. ■

We now compare the strong mixing coefficients and the tail functions of the regeneration times.

Proposition 9.7 *Let P be an irreducible and aperiodic stochastic kernel satisfying $\mathcal{M}(1, s, \nu)$. Let ψ be a function in Λ_0. Suppose that the stationary chain $(\xi_i)_{i \geq 0}$ with transition P and invariant law π satisfies $\sum_n \psi(n)\alpha_n < \infty$. Then, with the notation of Proposition 9.5, $\mathbb{E}_\pi(\psi^0(T_0)) < \infty$, and consequently $\sum_n \psi(n)\beta_n < \infty$.*

Remark 9.6 Proceeding as in Remark 9.5, one can prove that, if the above chain is geometrically strongly mixing, then the renewal times have a finite Laplace transform in a neighborhood of 0, and consequently the chain is geometrically β-mixing.

Application to arithmetic rates of mixing. Suppose that $\psi(n) = \max(2, n^{q-1})$ for some $q \geq 1$. By Proposition 9.7, $E_\pi(T_0^q) < \infty \Leftrightarrow \sum_{n>0} n^{q-1} \beta_n < \infty \Leftrightarrow \sum_{n>0} n^{q-1} \alpha_n < \infty$. Moreover, by (9.24), these conditions are equivalent to the condition $\mathbb{E}(\tau_0^{q+1}) < \infty$.

Next, using the fact that Proposition 9.7 holds for any function ψ in Λ_0, one can prove that, for any $q > 1$, $\mathbb{P}_\pi(T_0 > n) = O(n^{-q}) \Leftrightarrow \beta_n = O(n^{-q}) \Leftrightarrow \alpha_n = O(n^{-q})$. Moreover, by (9.24), these conditions are equivalent to the tail condition $\mathbb{P}(\tau_0 > n) = O(n^{-q-1})$.

Proof of Proposition 9.7 Let us apply (9.12) with $\lambda = \nu$. From the equality of masses, we have:

$$\pi Q^n(1) = \sum_{k=0}^{n-1} (\nu P^k(s) - \pi(s)) \nu Q^{n-k-1}(1) + \nu Q^n(1). \tag{9.28}$$

Now, by (9.4),

$$\pi(s) \nu Q^l(1) = \pi(s) \sum_{n \geq 0} \nu Q^{n+l}(s) = \mathbb{P}_\pi(T_0 = l + 1).$$

Hence (9.28) ensures that

$$\pi(s) \mathbb{P}_\pi(T_0 > n) = \sum_{k=0}^{n} (\nu P^k(s) - \pi(s)) \mathbb{P}_\pi(T_0 = n - k) + \mathbb{P}_\pi(T_0 = n + 1).$$

Now

$$\nu P^k(s) - \pi(s) = \mathbb{E}_\pi(s(\xi_{k+1}) | s(\xi_0) \geq U_0) - \mathbb{E}_\pi(s(\xi_{k+1})).$$

Therefore

$$\pi(s) |\nu P^k(s) - \pi(s)| \leq \tilde{\alpha}_{k+1} \leq \alpha_k. \tag{9.29}$$

Multiplying (9.29) by $\psi(n)$, summing on n and noting that, by Lemma 1 in Stone and Wainger (1967), $\psi(i + j) \leq \psi(i)\psi(j)$, we infer that

$$\sum_{n \geq 0} \mathbb{P}_\pi(T_0 = n) \psi^0(n) \leq (\pi(s))^{-2} \left(1 + \sum_{k \geq 0} \alpha_k \psi(k) \right) \sum_{n \geq 0} \mathbb{P}_\pi(T_0 = n) \psi(n). \tag{9.30}$$

For $M > 2$, let $\psi_M(n) = \psi(n) \wedge M$. Let us consider a function ψ in Λ_0 such that $\sum_n \psi(n) \alpha_n < \infty$. Set

$$C_\psi = (\pi(s))^{-2} \left(1 + \sum_{k \geq 0} \alpha_k \psi(k) \right).$$

By (9.30) applied to $\psi_M \leq \psi$, we get

$$\sum_{n \geq 0} \mathbb{P}_\pi(T_0 = n)\psi_M^0(n) \leq C_\psi \sum_{n \geq 0} \mathbb{P}_\pi(T_0 = n)\psi_M(n). \qquad (9.31)$$

We now prove that the series $\sum_n \mathbb{P}_\pi(T_0 = n)\psi(n)$ converges. Suppose that this series diverges. Then, for any positive n_0, the function g defined by

$$g(M) = \sum_{n \geq n_0} \mathbb{P}_\pi(T_0 = n)\psi_M(n)$$

is equivalent to $\sum_{n \geq 0} \mathbb{P}_\pi(T_0 = n)\psi_M(n)$ as M tends to infinity. Hence, by (9.31), for any positive integer n_0,

$$\limsup_{M \to +\infty} \frac{1}{g(M)} \sum_{n \geq n_0} \mathbb{P}_\pi(T_0 = n)\psi_M^0(n) \leq C_\psi. \qquad (9.32)$$

Now, by Lemma 2 in Stone and Wainger (1967), for any positive ε and any integer j_0, there exists a positive constant $c(\varepsilon, j_0)$ such that

$$\psi(n) \leq (1 + \varepsilon)\psi(n - j) + c(\varepsilon, j_0) \qquad (9.33)$$

for any $n > j_0$ and any $j \leq j_0$. Hence, for any positive j_0, there exists a rank n_0 such that, for any $n \geq n_0$ and any $j \leq j_0$, $2\psi(n - j) \geq \psi(n)$. This inequality still holds true for the function ψ_M. Hence $2\psi_M^0(n) \geq j_0\psi_M(n)$ for $n \geq n_0$. Consequently,

$$\frac{1}{g(M)} \sum_{n \geq n_0} \mathbb{P}_\pi(T_0 = n)\psi_M^0(n) \geq \frac{j_0}{2}.$$

For $j_0 \geq 4C_\psi$, the above inequality does not hold under (9.32) for M large. Hence the series $\sum_n \mathbb{P}_\pi(T_0 = n)\psi(n)$ converges. The second assertion follows from Corollary 9.1. ∎

Proceeding as in Exercise 6, Chap. 1, one can derive the corollary below from Proposition 9.7. In this corollary, the moment restriction on $f(\xi_0)$ comes from the fact that the functions ψ_0 defined in Definition 9.8 from a function ψ in Λ_0 fulfill the constraint $\log \psi_0(n) = o(n)$.

Corollary 9.2 *Let P be an irreducible and aperiodic stochastic kernel fulfilling $\mathcal{M}(1, s, \nu)$. Let π denote the unique invariant law and let $(\xi_i)_{i \geq 0}$ be the stationary Markov chain with transition P and initial law π. Then, for any numerical function f satisfying the integrability condition $\mathbb{E}(f^2(\xi_0) \log^+ |f(\xi_0)|) < \infty$, the integrals below are simultaneously convergent or divergent:*

$$\int_0^1 Q_{T_0}(u) Q_{f(\xi_0)}^2(u)du, \quad \int_0^1 \alpha^{-1}(u) Q_{f(\xi_0)}^2(u)du \quad \text{and} \quad \int_0^1 \beta^{-1}(u) Q_{f(\xi_0)}^2(u)du.$$

Corollary 9.2 proves that Theorem 4.2 can be applied in the case of Markov chains as soon as the random variable $Q_{T_0}(U)Q^2_{f(\xi_0)}(U)$ is integrable (here U denotes a random variable with uniform law over $[0, 1]$). In the forthcoming section, we prove that this condition cannot be improved.

9.7 On the Optimality of Theorem 4.2

In this section, we prove that condition (DMR) is sharp for the central limit theorem in the stationary case. In order to prove the optimality of this condition, we will construct a stationary Markov chain of random variables with values in $[0, 1]$ and strong mixing coefficients of the order of n^{-a} in such a way that, for any nonnegative decreasing function f such that the integrals appearing in Corollary 9.2 diverge, the random variable $\sum_{i=T_0}^{T_1-1} f(\xi_i)$ (the times T_0 and T_1 are defined in Sect. 9.3) fails to have a finite second moment. Applying then the converse of the central limit theorem, we will prove that the normalized and centered partial sums do not satisfy the central limit theorem. The chain will be defined from the transition probability kernel P introduced in Lemma 9.3. This transition can also be used to get lower bounds in the law of the iterated logarithm of Chap. 6 and lower bounds in the Marcinkiewicz–Zygmund type strong laws of Chap. 3 (we refer to Theorem 2 in Rio (1995a) for lower bounds in the strong laws). We mention that Bradley (1997) gives more general results with arbitrary rates of mixing.

Theorem 9.4 *For any real $a > 1$, there exists a stationary Markov chain $(U_i)_{i\in\mathbb{Z}}$ of random variables with uniform law over $[0, 1]$ and β-mixing coefficients $(\beta_n)_n$ such that:*

(i) $0 < \liminf_{n\to+\infty} n^a \beta_n \le \limsup_{n\to+\infty} n^a \beta_n < \infty,$
(ii) for any measurable and integrable function $f :]0, 1] \to \mathbb{R}$ such that

$$\int_0^1 u^{-1/a} f^2(u)du = +\infty, \qquad (a)$$

$n^{-1/2} \sum_{i=1}^n [f(U_i) - \mathbb{E}(f(U_i))]$ *does not converge in law.*

From Theorem 9.4 and Corollary 9.2, we get the following converse to Theorem 4.2 in Sect. 4.

Corollary 9.3 *Let $a > 1$ and let F be the distribution function of a centered and integrable random variable Z with atomless distribution. If*

$$\int_0^1 u^{-1/a} Q^2_Z(u)du = +\infty, \qquad (a)$$

then there exists a stationary Markov chain $(Z_i)_{i\in\mathbb{Z}}$ of random variables with distribution function F such that

(i) $0 < \liminf_{n \to +\infty} n^a \alpha_n \leq \limsup_{n \to +\infty} n^a \beta_n < \infty,$
(ii) $n^{-1/2} \sum_{i=1}^{n} Z_i$ *does not converge in distribution to a normal law.*

Proof of Theorem 9.4 Let $P(x, \cdot) = s(x)\nu + (1 - s(x))\delta_x$ be the transition probability kernel of Lemma 9.3. Take $\mathcal{X} =]0, 1]$ and $s(x) = x$. Let λ denote the Lebesgue measure on $[0, 1]$. For $a > 0$, let us define the regeneration measure ν by $\nu = (1 + a)x^a \lambda$. Then the chain is positively recurrent and the invariant law π is given by $\pi = ax^{a-1}\lambda$ (see the proof of Lemma 9.3). Let $t(x) = 1 - x$. Then for any positive k

$$\mathbb{P}_\pi(\tau > k) = \mathbb{E}_\pi(t^k) = k^{-a} \int_0^k (1 - x/k)^k a x^{a-1} dx. \tag{9.34}$$

Consequently,

$$\lim_{k \to +\infty} k^a \mathbb{E}_\pi(t^k) = a\Gamma(a), \tag{9.35}$$

where Γ is Euler's gamma function. Since the distribution function of π is $F_\pi(x) = x^a$, the stationary sequence $(U_i)_i$ defined by $U_i = \xi_i^a$ is a stationary Markov chain of random variables with uniform law over $[0, 1]$. This chain has the same β-mixing coefficients as the initial chain $(\xi_i)_{i \in \mathbb{Z}}$. Now Theorem 9.4(i) follows from Propositions 9.4 and (9.35).

We now prove (ii). We may assume, without loss of generality, that $\mathbb{E}(f(U_i)) = 0$. Now, using the renewal scheme, we prove that some compound sums defined from the variables $f(U_i)$ are partial sums of independent and identically distributed random variables. With the notation of Definition 9.5,

$$\sum_{i=1}^{T_n-1} f(U_i) = \sum_{i=1}^{\tau-1} f(U_i) + \sum_{k=0}^{n-1} \tau_k f(U_{T_k}). \tag{9.36}$$

We now prove that

$$\sum_{i=1}^{T_n-1} f(U_i) - \sum_{i=1}^{[n\mathbb{E}(\tau_1)]} f(U_i) = o_P(\sqrt{n}). \tag{9.37}$$

To prove (9.37), we start by noting that the random variables $(X_{T_k}, \tau_k)_{k>0}$ are independent and identically distributed. Let $\zeta_k = X_{T_k}$. The random variables ζ_k are independent with common law ν. Now

$$\mathbb{P}(\tau_k > n | \zeta_k = \zeta) = (1 - \zeta)^n, \tag{9.38}$$

which ensures that $\mathbb{E}(\tau_1^2) < \infty$ for any $a > 1$. Hence, by the usual central limit theorem, $(T_n - n\mathbb{E}(\tau_1))/\sqrt{n}$ converges in distribution to a nondegenerate normal law. It follows that, for any positive ϵ, there exists some positive A such that

$$\liminf_{n \to +\infty} \mathbb{P}(n\mathbb{E}(\tau_1) \in [T_{[n-A\sqrt{n}]}, T_{[n+A\sqrt{n}]}]) \geq 1 - \epsilon. \tag{9.39}$$

Now, by (9.38),

$$\mathbb{E}|\tau_k f(U_{T_k})| = \int_0^1 |f(\zeta^a)|\alpha\zeta^{a-1}d\zeta < \infty \quad \text{and} \quad \mathbb{E}(\tau_k f(U_{T_k})) = 0.$$

Therefore, applying the strong law of large numbers to the sequence $(\tau_k f(U_{T_k}))_{k>0}$, we get that

$$n^{-1/2} \sup_{m\in[n-A\sqrt{n},n+A\sqrt{n}]} \left| \sum_{k=1}^{n} \mathbb{E}(\tau_k f(U_{T_k})) - \sum_{k=1}^{m} \mathbb{E}(\tau_k f(U_{T_k})) \right| \longrightarrow_{P} 0 \text{ as } n \to \infty.$$

Now, from (9.36), the random variable $n^{-1/2}|\sum_{i=1}^{T_n-1} f(U_i) - \sum_{i=1}^{[n\mathbb{E}(\tau_1)]} f(U_i)|$ is less than the above random variable on the event $(n\mathbb{E}(\tau_1) \in [T_{[n-A\sqrt{n}]}, T_{[n+A\sqrt{n}]}])$. Therefore, putting together (9.38) and the above inequality we get (9.37).

From (9.37), if the compound sums $\Delta_n = n^{-1/2} \sum_{k=0}^{n-1} \tau_k f(U_{T_k})$ do not converge in distribution to a normal law as n tends to ∞, then the normalized sums $n^{-1/2} \sum_{i=1}^{n} f(U_i)$ do not satisfy the central limit theorem. Now by the converse of the central limit theorem (see Feller (1950) for more about this), Δ_n converges in law to a normal random variable if and only if $\mathbb{E}(\tau_k^2 f^2(U_{T_k})) < \infty$. By (9.38), this condition holds if and only if

$$\mathbb{E}(\zeta_1^{-2}[f(\zeta_1^a)]^2) = (1+a) \int_0^1 \zeta^{a-2}[f(\zeta^a)]^2 d\zeta < \infty.$$

Setting $u = \zeta^a$ in the above integral, we then get Theorem 9.4(ii), which completes the proof. ∎

Exercises

(1) Let $p > 2$. Prove that, for any $a > 1$ and any continuous distribution function F such that $\int_{\mathbb{R}} |x|^p dF(x) < \infty$ and $\int_{\mathbb{R}} x dF(x) = 0$, there exists a stationary sequence Z of random variables with common law F and β-mixing coefficients β_i of the order of i^{-a} such that

$$\mathbb{E}(|S_n|^p) \geq cn^p \int_0^{n^{-a}} Q_0^p(u)du,$$

for some positive constant c. Compare this result with Theorem 6.3 and Corollary 6.1.

(2) Let $(\xi_i)_{i\in\mathbb{Z}}$ be a stationary Markov chain. Assume that the uniform mixing coefficients φ_n converge to 0 as n tends to ∞. Prove that $\varphi_n = O(\rho^n)$ for some ρ in $[0, 1[$.

Annex A
Young Duality and Orlicz Spaces

In this annex we recall some basic properties of the Young transform of convex functions. Next we define the Orlicz spaces and the Orlicz norms and we give elementary applications of these notions.

Let us introduce the class of convex functions

$$\bar{\Phi} = \{\phi : \mathbb{R}^+ \to \bar{\mathbb{R}}^+ : \phi \text{ convex, nondecreasing, left-continuous, } \phi(0) = 0\}.$$

We denote by D_ϕ the set of nonnegative reals x such that $\phi(x) < \infty$. From the convexity of ϕ, the set D_ϕ is an interval.

A.1. Young Duality

For ϕ in $\bar{\Phi}$, let

$$G_\phi = \{(x, y) \in D_\phi \times \mathbb{R}^+ \text{ such that } y > \phi(x)\}$$

denote the super-graph of ϕ and let $\bar{G}_\phi$ denote the closure of G_ϕ. The Young dual function of ϕ is defined by

$$\phi^*(\lambda) = \sup_{x \in D_\phi} (\lambda x - \phi(x)) \text{ for any } \lambda \geq 0.$$

Thus $z = \phi^*(\lambda)$ if and only if the straight line with equation $y = \lambda x - z$ is tangent to G_ϕ. In that case, $D_{\lambda,z} \cap \bar{G}_\phi \neq \emptyset$ and $D_{\lambda,z} \cap G_\phi = \emptyset$. It follows that

$$z \geq \phi^*(\lambda) \text{ if and only if } D_{\lambda,z} \cap G_\phi = \emptyset. \tag{A.1}$$

Starting from (A.1), we now prove that ϕ^* belongs to $\bar{\Phi}$. Clearly ϕ^* is nondecreasing. Noticing that the straight line $D_{0,z}$ with equation $y = -z$ does not intersect G_ϕ

© Springer-Verlag GmbH Germany 2017
E. Rio, *Asymptotic Theory of Weakly Dependent Random Processes,*
Probability Theory and Stochastic Modelling 80,
DOI 10.1007/978-3-662-54323-8

if and only if $z \geq 0$, we get that $\phi^*(0) = 0$. To prove that ϕ^* is a convex function, we will apply the elementary lemma below (proof omitted).

Lemma A.1 *For any set I and any collection $(\psi_i)_{i \in I}$ of convex functions, $\psi = \sup_{i \in I} \psi_i$ is a convex function.*

The convexity of ϕ^* follows immediately by applying Lemma A.1 to the collection of functions $(\psi_x)_{x \in D_\phi}$ defined by $\psi_x(\lambda) = \lambda x - \phi(x)$. We now prove that ϕ^* is left-continuous. First note that

$$\phi(x) + \phi^*(\lambda) \geq \lambda x \text{ for any } (x, \lambda) \in D_\phi \times D_{\phi^*}. \tag{A.2}$$

This inequality is called Young's inequality. If $l = \lim_{\lambda \nearrow \lambda_0} \phi^*(\lambda)$ is finite, then, taking the limit as λ tends to λ_0 in Young's inequality, we get that $\phi(x) + l \geq \lambda_0 x$ for any x in D_ϕ, which ensures that $\phi(\lambda_0)$ is finite and satisfies $\phi(\lambda_0) \leq l$. Since ϕ^* is nondecreasing, it follows that $l = \phi(\lambda_0)$.

We now prove that $\phi^{**} = \phi$. From Young's inequality $\phi(x) \geq \phi^{**}(x)$. Suppose now that $y > \phi^{**}(x)$. Then, for any nonnegative λ, $y > \lambda x - \phi^*(\lambda)$. Now, from the convexity of ϕ, G_ϕ is the intersection of all the half-planes $y > \lambda x - \phi^*(\lambda)$. Hence $y > \phi(x)$, which proves the converse inequality $\phi(x) \leq \phi^{**}(x)$.

Derivatives of ϕ^*. The derivatives of ϕ^* satisfy the relations below:

$$(\phi^*)'(\lambda + 0) = \phi'^{-1}(\lambda + 0) \text{ and } (\phi^*)'(\lambda - 0) = \phi'^{-1}(\lambda - 0) = \phi'^{-1}(\lambda). \tag{A.3}$$

To prove (A.3), we consider the intersection points of the straight line $y = \lambda x - \phi^*(\lambda)$ with $\bar{G}_\phi$. Since the inverse functions are left-continuous the intersection point $(x(\lambda), y(\lambda))$ with maximal coordinate x satisfies $x(\lambda) = \phi'^{-1}(\lambda + 0)$. For arbitrary $\varepsilon > 0$, let us consider the straight line with equation $y - \phi(x(\lambda)) = (\lambda + \varepsilon)(x - x(\lambda))$. For $x = 0$ in this equation, $y = \phi^*(\lambda) + \varepsilon x(\lambda)$. Consequently, $\phi^*(\lambda + \varepsilon) \geq \phi^*(\lambda) + \varepsilon x(\lambda)$. Next, for any $x > x(\lambda)$, $\phi'(x) > \lambda$. Hence, for $x > x(\lambda)$ and ε small enough, $\phi'(x) \geq \lambda + \varepsilon$. Therefore, for any $t \geq x$, $(\lambda + \varepsilon)t - \phi(t) \leq (\lambda + \varepsilon)x - \phi(x)$. Now, for any $t \leq x$,

$$(\lambda + \varepsilon)t - (\phi^*(\lambda) + \varepsilon x) \leq \lambda t - \phi^*(\lambda) \leq \phi(t).$$

Both the above two inequalities ensure that $(\lambda + \varepsilon)t - (\phi^*(\lambda) + \varepsilon x) \leq \phi(t)$ for any positive t. Hence $\phi^*(\lambda + \varepsilon) \leq \phi^*(\lambda) + \varepsilon x$. Thus we have proved that

$$\phi^*(\lambda) + \varepsilon x(\lambda) \leq \phi^*(\lambda + \varepsilon) \leq \phi^*(\lambda) + \varepsilon x. \tag{A.4}$$

The left-hand inequality in (A.3) follows immediately from (A.4). The proof of the second part of (A.3) is similar.

Inverse function of ϕ^*. The lemma below furnishes a direct way to compute the inverse function of ϕ^*.

Lemma A.2 *For any ϕ in $\bar{\Phi}$ and any positive x,*

$$\phi^{*-1}(x) = \inf_{t \in D_\phi} t^{-1}(\phi(t) + x).$$

Proof of Lemma A.2 The slope of the straight line $D_{x,t}$ containing $(0, -x)$ and $(t, \phi(t))$ is equal to $t^{-1}(\phi(t) + x)$. Let t_0 be the point which realizes the minimum of this slope and λ_0 be the corresponding slope. Then the straight line D_{x,t_0} is tangent to the curve $y = \phi(t)$. Consequently, $\phi(\lambda_0) = x$, which completes the proof of Lemma A.2.

A.2. Orlicz Spaces

Let ϕ be any function in $\bar{\Phi}$ such that $\phi \neq 0$. For any random vector Z in a normed vector space $(E, |\cdot|)$, the Luxemburg norm associated to ϕ is defined by

$$\|Z\|_\phi = \inf\{c > 0 : \mathbb{E}(\phi(|Z|/c) \leq 1\} \tag{A.5}$$

if there exists some positive real c such that $\mathbb{E}(\phi(|Z|/c)) < \infty$, and by $\|Z\|_\phi = +\infty$ otherwise.

We now prove that $\|\cdot\|_\phi$ is a norm. Clearly $\|\lambda Z\|_\phi = |\lambda| \|\lambda Z\|_\phi$. Next, from the convexity of ϕ, for $c > \|Z\|_\phi$ and $c' > \|Z'\|_\phi$,

$$\mathbb{E}\left(\phi(|Z+Z'|/(c+c'))\right) \leq \frac{c}{c+c'}\mathbb{E}(\phi(|Z|/c)) + \frac{c'}{c+c'}\mathbb{E}(\phi(|Z'|/c')) \leq 1, \tag{A.6}$$

which proves the triangle inequality. Now, if $\|Z\|_\phi = 0$, then, for any positive integer n, $\phi(n|Z|) = 0$ almost surely. Consequently, for any positive a such that $\phi(a) > 0$, $n|Z| \leq a$ almost surely, which implies that $Z = 0$ almost surely, which completes the proof. Throughout the sequel, we denote by L^ϕ the normed space of real-valued random variables Z such that $\|Z\|_\phi < \infty$.

We now give classical extensions of the Hölder inequalities to Orlicz spaces. Let X and Y be nonnegative random variables. Then, by Young's inequality (A.2),

$$\mathbb{E}(XY) \leq \mathbb{E}(\phi(X) + \phi^*(Y)). \tag{A.7}$$

Now, let $c > \|X\|_\phi$ and $c' > \|Y\|_{\phi^*}$. Applying (A.7) to $(X/c, Y/c')$, we get that $\mathbb{E}(XY) \leq 2cc'$. It follows that

$$\mathbb{E}(XY) \leq 2\|X\|_\phi \|Y\|_{\phi^*} \text{ for any } X \in L^\phi \text{ and any } Y \in L^{\phi^*}. \tag{A.8}$$

We now give some applications of these theoretical results to particular functions. We refer to Dellacherie and Meyer (1975) for more about the theoretical results.

A.3. Applications to Classical Orlicz Spaces

First take $p > 1$ and $\phi(x) = (x^p/p)$. Let $q = p/(p-1)$ be the conjugate exponent. Then $\phi^*(y) = q^{-1}y^q$. Hence, by (A.7),

$$\mathbb{E}(XY) \leq \frac{1}{p}\mathbb{E}(X^p) + \frac{1}{q}\mathbb{E}(Y^q). \tag{A.9}$$

Applying this inequality to $X/\|X\|_p$ and $Y/\|Y\|_q$, we get the usual Hölder inequality

$$\mathbb{E}(XY) \leq (\mathbb{E}(X^p))^{1/p}(\mathbb{E}(Y^q))^{1/q}. \tag{A.10}$$

Note that (A.8) implies (A.10) only in the case $p = q = 2$. For $p \neq 2$, a direct application of (A.8) leads to the multiplicative loss $2p^{-1/p}q^{-1/q}$.

Now, let $\phi(x) = e^x - 1 - x$. Then the equation of the tangent to the curve $(x, \phi(x))$ at the point $(t, \phi(t))$ is $y - \phi(t) = (x - t)(e^t - 1)$, whence $\phi^*(e^t - 1) = (t-1)e^t + 1$. Now, if $\lambda = e^t - 1$, then $t = \log(1 + \lambda)$ and consequently

$$\phi^*(\lambda) = (1 + \lambda)(\log(1 + \lambda) - 1) + 1 = (1 + \lambda)\log(1 + \lambda) - \lambda. \tag{A.11}$$

Affine transformations. Let A be defined by $A(x, y) = (ax, by + cx)$, with $a > 0$, $b > 0$ and $c \geq 0$. Let ϕ_A be the map whose graph is the image under A of the graph of ϕ. Then

$$\phi_A(x) = b\phi(x/a) + cx/a. \tag{A.12}$$

Since the tangent to G_ϕ with slope λ is changed to the tangent to G_{ϕ_A} with slope $(b\lambda + c)/a$ by the map A, we get that

$$\phi_A^*(\lambda') = b\phi^*((a\lambda' - c)/b) \text{ for any } \lambda' \geq c/a \text{ and } \phi_A^*(\lambda') = 0 \text{ otherwise.} \tag{A.13}$$

Annex B
Exponential Inequalities for Sums of Independent Random Variables

This annex is devoted to some common exponential inequalities for sums. We refer to Chap. 2 in Bercu et al. (2015) for more about this subject. Throughout the section $Z_1, Z_2, \ldots$ is a sequence of independent real-valued random variables with finite variance. Set

$$S_0 = 0, \quad S_k = (Z_1 - \mathbb{E}(Z_1)) + \cdots + (Z_k - \mathbb{E}(Z_k)) \quad \text{and} \quad S_n^* = \max(S_0, S_1, \ldots, S_n).$$

We start by recalling a version of Bennett's inequality due to Fuk and Nagaev (1971).

Theorem B.1 *Let K be a positive constant. Assume that $Z_1, Z_2, \ldots$ satisfy the additional conditions $Z_i \leq K$ almost surely. Then, for any $V \geq \mathbb{E}(Z_1^2) + \cdots + \mathbb{E}(Z_n^2)$ and any positive λ,*

$$\mathbb{P}(S_n^* \geq \lambda) \leq \exp(-K^{-2}Vh(\lambda K/V)), \tag{a}$$

with $h(x) = (1 + x)\log(1 + x) - x$. If, furthermore, $|Z_i| \leq K$ almost surely, then

$$\mathbb{P}(\sup_{k \in [1,n]} |S_k| \geq \lambda) \leq 2\exp(-K^{-2}Vh(\lambda K/V)). \tag{b}$$

Proof The proof is based on the classical Crámer–Chernoff calculation, which we now recall in Lemma B.1 below.

Lemma B.1 *Let γ be a nondecreasing convex function on $\mathbb{R}^+$ such that $\gamma(t) \geq \log \mathbb{E}(\exp(tS_n))$ for any nonnegative t. Then, for any positive λ,*

$$\log(\mathbb{P}(S_n^* \geq \lambda)) \leq \inf_{t>0}(\gamma(t) - t\lambda) = -\gamma^*(\lambda).$$

Proof of Lemma B.1 For t in the domain of γ, set $M_k(t) = \exp(tS_k)$. Then, from Jensen's inequality, $(M_k(t))_{k\geq 0}$ is a nonnegative submartingale. Hence, by Doob's

© Springer-Verlag GmbH Germany 2017
E. Rio, *Asymptotic Theory of Weakly Dependent Random Processes*,
Probability Theory and Stochastic Modelling 80,
DOI 10.1007/978-3-662-54323-8

maximal inequality,

$$\mathbb{P}(S_n^* \geq \lambda) \leq \mathbb{E}(\exp(t S_n - t\lambda)) \leq \exp(\gamma(t) - t\lambda). \tag{B.1}$$

Lemma B.1 follows immediately. ∎

Proof of Theorem B.1 From Lemma B.1, it is enough to prove that

$$\log \mathbb{E}(\exp(t S_n)) \leq K^{-2} V(\exp(t K) - t K - 1), \tag{B.2}$$

and next to apply (A.11) and (A.13). Now, using the independence of the random variables Z_i and then the concavity of the logarithm, we get that

$$\log \mathbb{E}(\exp(t S_n)) = \sum_{i=1}^{n} (\log \mathbb{E}(\exp(t Z_i)) - t\mathbb{E}(Z_i))$$

$$\leq \sum_{i=1}^{n} \mathbb{E}(\exp(t Z_i) - t Z_i - 1). \tag{B.3}$$

Next, the function ψ defined by $\psi(0) = 1/2$ and $\psi(x) = x^{-2}(e^x - x - 1)$ for $x \neq 0$ is nondecreasing. Since $Z_i \leq K$ almost surely, it follows that

$$\mathbb{E}(\exp(t Z_i) - t Z_i - 1) \leq K^{-2}\mathbb{E}(Z_i^2)(\exp(t K) - t K - 1).$$

Combining this inequality with (B.3), we get (B.2). Hence (a) holds. To prove (b), apply (a) to the random variables $-Z_1, \ldots, -Z_n$ and add the two inequalities. ∎

Below we give a one-sided version of the Bernstein inequality, which allows us to consider random variables with finite Laplace transform only in a right neighborhood of the origin. We refer to Pollard (1984) for the usual Bernstein inequality.

Theorem B.2 *Let $Z_1, \ldots, Z_n$ be a finite sequence of independent random variables. Set $Z_{i+} = \max(Z_i, 0)$. Suppose that there exist positive constants K and V such that*

$$\sum_{i=1}^{n} \mathbb{E}(Z_i^2) \leq V \text{ and } \sum_{i=1}^{n} \frac{1}{m!}\mathbb{E}(Z_{i+}^m) \leq \frac{1}{2}V K^{m-2} \text{ for any } m \geq 3. \tag{a}$$

Then, for any positive λ,

$$\mathbb{P}(S_n^* \geq \lambda) \leq \exp(-z) \leq \exp\left(-\frac{\lambda^2}{2(V + K\lambda)}\right),$$

where $z = z(\lambda)$ is the positive real defined by $Kz + \sqrt{2Vz} = \lambda$.

Proof If $\lambda = Kz + \sqrt{2Vz}$ then $\lambda^2 \leq 2(V + K\lambda)z$, which implies the second inequality. We now prove the first inequality. Starting from (B.3) and noting that $e^x - 1 - x \leq x^2/2$ for any negative x, we get that

$$\log \mathbb{E}(\exp(t S_n)) \le \frac{t^2}{2} \sum_{i=1}^{n} \mathbb{E}(Z_i^2) + \sum_{i=1}^{n} \sum_{m=3}^{\infty} \frac{t^m}{m!} \mathbb{E}(Z_{i+}^m).$$

Therefore, if assumption (a) holds, then, for any nonnegative t,

$$\log \mathbb{E}(\exp(t S_n)) \le \gamma(t) = \frac{1}{2} V t^2 / (1 - Kt). \tag{B.4}$$

From Lemma B.1, the proof of Theorem B.2 will be complete if we prove that

$$\gamma^{*-1}(z) = Kz + \sqrt{2Vz}. \tag{B.5}$$

To prove (B.5), we apply Lemma A.2: setting $u = (1/t) - K$ in the formula of Lemma A.2, we get that

$$\gamma^{*-1}(z) = \inf_{t \in]0,1/K[} (Vt/(1 - Kt) + (z/t)) = \inf_{u>0} ((V/u) + uz + Kz) = \sqrt{2Vz} + Kz,$$

which proves (B.5). Hence Theorem B.2 holds. ∎

We now give an application of Theorem B.2 to bounded random variables in Corollary B.1 below, which improves on the usual Bernstein inequality.

Corollary B.1 *Let $Z_1, \ldots, Z_n$ be a finite sequence of independent random variables. Suppose that $Z_i \le M$ almost surely for any i. Set*

$$D_n = \sum_{i=1}^{n} \mathbb{E}(Z_i^2) \quad and \quad L_n = (M D_n)^{-1} \sum_{i=1}^{n} \mathbb{E}(Z_{i+}^3).$$

Then, for any positive x,

$$\mathbb{P}(S_n^* \ge \sqrt{2 D_n x} + \max(L_n/3, 1/4) M x) \le \exp(-x).$$

Remark B.1 The usual multiplicative factor before Mx is $1/3$. Since $L_n \le 1$, Corollary B.1 gives slightly better bounds.

Proof of Corollary B.1 For any $m \ge 4$, since $Z_{i+}^m \le M^{m-3} Z_{i+}^3$,

$$\sum_{i=1}^{n} \frac{1}{m!} \mathbb{E}(Z_{i+}^m) \le \frac{M^{m-3}}{m!} \sum_{i=1}^{n} \mathbb{E}(Z_{i+}^3) \le \frac{M^{m-2} L_n}{m!} D_n \le \left(\frac{M}{4}\right)^{m-3} \frac{M L_n}{3} D_n.$$

Now $(M/4)^{m-3}(ML_n/3) \leq (\max(M/4, ML_n/3))^{m-2}$. Consequently, assumption (a) of Theorem B.2 holds true with $V = D_n$ and $K = \max(M/4, ML_n/3)$, which completes the proof of Corollary B.1. ∎

We now give some applications of Theorem B.1 to deviation inequalities for sums of unbounded random variables. The inequalities below are due to Fuk and Nagaev (1971).

Theorem B.3 *Let $Z_1, \ldots, Z_n$ be a finite sequence of independent and square integrable random variables. Then, for any $V \geq \sum_{i=1}^n \mathbb{E}(Z_i^2)$ and any pair (λ, x) of strictly positive reals,*

$$\mathbb{P}(S_n^* \geq \lambda) \leq \exp(-x^{-2}Vh(\lambda x/V)) + \sum_{i=1}^n \mathbb{P}(Z_i > x), \qquad (a)$$

with $h(x) = (1+x)\log(1+x) - x$. Moreover, for any positive ε,

$$\mathbb{P}(S_n^* \geq (1+\varepsilon)\lambda) \leq \exp(-x^{-2}Vh(\lambda x/V)) + \frac{1}{\lambda\varepsilon}\sum_{i=1}^n \mathbb{E}((Z_i - x)_+). \qquad (b)$$

Proof Set

$$\bar{Z}_i = Z_i \wedge x, \ \ \bar{S}_k = \sum_{i=1}^k (\bar{Z}_i - \mathbb{E}(\bar{Z}_i)) \ \text{ and } \ \bar{S}_n^* = \sup_{k \in [0,n]} \bar{S}_k,$$

with the convention that $\bar{S}_0 = 0$. Since

$$S_k \leq \bar{S}_k + \sum_{i=1}^k (Z_i - x)_+ - \sum_{i=1}^k \mathbb{E}(Z_i - x)_+ \leq \bar{S}_k + \sum_{i=1}^n (Z_i - x)_+,$$

we have:

$$S_n^* \leq \bar{S}_n^* + \sum_{i=1}^n (Z_i - x)_+. \qquad (B.6)$$

Let us prove (a). If $Z_i \leq x$ for any i in $[1, n]$, then, by (B.6), $S_n^* \leq \bar{S}_n^*$. It follows that

$$\mathbb{P}(S_n^* > \bar{S}_n^*) \leq \mathbb{P}(Z_1 > x) + \cdots + \mathbb{P}(Z_n > x).$$

Therefore

$$\mathbb{P}(S_n^* \geq \lambda) \leq \mathbb{P}(\bar{S}_n^* \geq \lambda) + \sum_{i=1}^{n} \mathbb{P}(Z_i > x)$$

$$\leq \exp(-x^{-2} V h(\lambda x / V)) + \sum_{i=1}^{n} \mathbb{P}(Z_i > x) \tag{B.7}$$

by Theorem B.1(a), since $\mathbb{E}(\bar{Z}_i^2) \leq \mathbb{E}(Z_i^2)$.

We now prove (b). Applying (B.6), we obtain that

$$\mathbb{P}(S_n^* \geq (1+\varepsilon)\lambda) \leq \mathbb{P}(\bar{S}_n^* \geq \lambda) + \mathbb{P}\left(\sum_{i=1}^{n} (Z_i - x)_+ \geq \varepsilon\lambda\right)$$

$$\leq \mathbb{P}(\bar{S}_n^* \geq \lambda) + (\varepsilon\lambda)^{-1} \sum_{i=1}^{n} \mathbb{E}(Z_i - x)_+$$

by the Markov inequality applied to the second term on the right-hand side. The end of the proof is then exactly the same as the end of the proof of (B.7). ∎

We now state an inequality of Hoeffding for independent and bounded random variables.

Theorem B.4 *Let $Z_1, \ldots, Z_n$ be a finite sequence of independent bounded real-valued random variables. Suppose that $Z_i^2 \leq M_i$ for any i in $[1, n]$. Then, for any positive λ,*

$$\mathbb{P}(S_n^* \geq \lambda) \leq \exp\left(-x^2/(2M_1 + \cdots + 2M_n)\right). \tag{a}$$

Moreover,

$$\mathbb{P}(\sup_{k \in [1,n]} |S_k| \geq \lambda) \leq 2\exp\left(-x^2/(2M_1 + \cdots + 2M_n)\right). \tag{b}$$

Proof It is enough to prove that, if Z belongs to $[-m, m]$ almost surely, then

$$\mathbb{E}(\exp(tZ - t\mathbb{E}(Z))) \leq \exp(t^2 m^2 / 2), \tag{B.8}$$

which ensures that $\log \mathbb{E} \exp(t S_n) \leq t^2 (M_1 + \cdots + M_n)/2$, and next to apply Lemma B.1. To prove (B.8), we may assume that $m = 1$. From the convexity of the exponential function,

$$2\exp(tZ) \leq (1 - Z)\exp(-t) + (1 + Z)\exp(t).$$

Set $q = \mathbb{E}(Z)$. Taking the expectation of the above inequality, we get that

$$\mathbb{E}(\exp(tZ)) \leq \cosh t + q \sinh t.$$

Let $f(t) = \cosh t + q \sinh t$. Taking the logarithm of the above inequality, we have:

$$\log \mathbb{E}(\exp(tZ - t\mathbb{E}(Z)) \leq \log f(t) - qt.$$

Now $(\log f)'' = (f''/f) - (f'/f)^2 \leq 1$, since $f'' = f$. Hence, integrating this differential inequality twice, we get that $\log f(t) \leq qt + t^2/2$, which implies Theorem B.4(a). Theorem 1.4(b) is obvious. ∎

$^{(*)}$ **Sums of non-independent random variables.** Consider now a random variable which is equal to $A + B$, where A and B are real-valued random variables, and B may depend on A. Suppose that the Laplace transforms of A and B are finite on a right neighborhood of 0 and let γ_A and γ_B denote the log-Laplace transforms of A and B, respectively. Adding Chernoff's deviation inequalities yields

$$\mathbb{P}(A + B \geq \gamma_A^{*-1}(z) + \gamma_B^{*-1}(z)) \leq 2\exp(-z).$$

In fact, the above inequality can be improved by a factor of 2, as proved by the lemma below, stated by Rio (1994). The original proof in Rio (1994) was due to Jean Bretagnolle. Here we will give a shorter proof based on Lemma A.2.

Lemma B.2 *Let A and B be real-valued and centered random variables with respective log-Laplace transforms γ_A and γ_B. Suppose that γ_A and γ_B are finite in a right neighborhood of 0. Then, for any positive z,*

$$\gamma_{A+B}^{*-1}(z) \leq \gamma_A^{*-1}(z) + \gamma_B^{*-1}(z). \tag{a}$$

Consequently, for any positive z,

$$\mathbb{P}(A + B > \gamma_A^{*-1}(z) + \gamma_B^{*-1}(z)) \leq \exp(-z). \tag{b}$$

Remark B.2 Clearly (a) may be extended to a finite sum of random variables. For example, suppose that $A_1, A_2, \ldots A_n$ is a finite collection of random variables satisfying

$$\log \mathbb{E}(\exp(tA_i)) \leq \sigma_i^2 t^2/(1 - c_i t) \text{ for } t \in [0, c_i[, \text{ with } c_i > 0 \text{ and } \sigma_i > 0.$$

Then Inequality (B.5) together with Lemma B.2 yield

$$\mathbb{P}(A_1 + A_2 + \cdots + A_n \geq \sigma\sqrt{2z} + cz) \leq \exp(-z), \tag{B.9}$$

with $\sigma = \sigma_1 + \sigma_2 + \cdots + \sigma_n$ and $c = c_1 + c_2 + \cdots + c_n$.

Proof of Lemma B.2 By Hölder's inequality, for any reals $p > 1$ and $q > 1$ with $(1/p) + (1/q) = 1$,

$$\log \mathbb{E}(\exp(t(A + B))) = \gamma_{A+B}(t) \leq p^{-1}\gamma_A(tp) + q^{-1}\gamma_B(tq).$$

Applying Lemma A.2, we infer that

$$\gamma_{A+B}^{*-1}(z) \le \inf_{p>1} \inf_{t>0} \left(\frac{\gamma_A(tp) + z}{tp} + \frac{\gamma_B(tp/(p-1)) + z}{tp/(p-1)} \right).$$

Now, the map $(t, p) \to (tp, tp/(p-1))$ is a diffeomorphism from $\mathbb{R}_+^* \times]1, \infty[$ onto $\mathbb{R}_+^{*2}$. It follows that the term on the right-hand side of the above inequality is equal to $\gamma_A^{*-1}(z) + \gamma_B^{*-1}(z)$, which completes the proof of Lemma B.2(a). Part (b) is a direct consequence of (a).

Annex C
Upper Bounds for the Weighted Moments

In this annex, we give upper bounds for the quantities $M_{p,\alpha}(Q)$ introduced in chapters one to six. Throughout Annex C, let Q be the quantile function of a nonnegative random variable X. For $p \geq 1$, let

$$M_{p,\alpha}(Q) = \int_0^1 [\alpha^{-1}(u)]^{p-1} Q^p(u) du, \quad M_{p,\alpha,n}(Q) = \int_0^1 [\alpha^{-1}(u) \wedge n]^{p-1} Q^p(u) du. \tag{C.1}$$

Here we give sufficient conditions ensuring that $M_{p,\alpha}(Q)$ is finite. We also give some precise upper bounds on $M_{p,\alpha}(Q)$ and $M_{p,\alpha,n}(Q)$ depending on the mixing rate and the quantile function Q or the tail function of X.

We first bound $M_{p,\alpha}(Q)$ under moment conditions on X. Let U be a random variable with the uniform distribution over $[0, 1]$. Then X and $Q(U)$ are identically distributed. Hence

$$\mathbb{E}(X^r) = \int_0^1 Q^r(u) du \text{ for any } r > 0. \tag{C.2}$$

Suppose now that $\mathbb{E}(X^r) < \infty$ for some $r > 1$. Then, for any p in $]1, r[$, by Hölder's inequality applied with exponents $r/(r - p)$ and r/p, we get that

$$M_{p,\alpha}(Q) \leq \left(\int_0^1 [\alpha^{-1}(u)]^{(p-1)r/(r-p)} du \right)^{1-p/r} \left(\int_0^1 Q^r(u) du \right)^{p/r}. \tag{C.3}$$

Now, proceeding as in the proof of (1.25), we note that, for any positive q,

$$\int_0^1 [\alpha^{-1}(u)]^q du = \sum_{i \geq 0} ((i + 1)^q - i^q) \alpha_i. \tag{C.4}$$

© Springer-Verlag GmbH Germany 2017
E. Rio, *Asymptotic Theory of Weakly Dependent Random Processes*,
Probability Theory and Stochastic Modelling 80,
DOI 10.1007/978-3-662-54323-8

Next $(i + 1)^q - i^q \leq \max(q, 1)(i + 1)^{q-1}$, and consequently

$$\int_0^1 [\alpha^{-1}(u)]^q du \leq \max(q, 1) \sum_{i \geq 0} (i + 1)^{q-1} \alpha_i. \tag{C.5}$$

Both (C.3) and (C.5) ensure that

$$M_{p,\alpha}(Q) \leq \max(1, (p - 1)^{1-p/r} e^{p/r}) \|X\|_r^p \left(\sum_{i \geq 0} (i + 1)^{(pr-2r+p)/(r-p)} \alpha_i \right)^{1-p/r}. \tag{C.6}$$

If the random variable X is bounded, then, taking $r = \infty$ in (C.6), we get that

$$M_{p,\alpha}(Q) \leq \max(1, p - 1) \|X\|_\infty^p \sum_{i \geq 0} (i + 1)^{p-2} \alpha_i. \tag{C.7}$$

Consequently, $M_{p,\alpha}(Q)$ is finite as soon as there exists some real $r > p$ such that

$$\mathbb{E}(X^r) < \infty \quad \text{and} \quad \sum_{i \geq 0} (i + 1)^{(pr-2r+p)/(r-p)} \alpha_i < \infty. \tag{C.8}$$

We now bound the quantities $M_{p,\alpha}(Q)$ and $M_{p,\alpha,n}(Q)$ in a slightly different way. Clearly

$$[\alpha^{-1}(u)]^{p-1} = (p - 1) \int_0^\infty \mathbb{1}_{u < \alpha(t)} t^{p-2} dt.$$

Hence, by the Fubini–Tonelli theorem,

$$M_{p,\alpha}(Q) = (p - 1) \int_0^\infty t^{p-2} \left(\int_0^{\alpha(t)} Q^p(u) du \right) dt.$$

Next

$$(p - 1) \int_i^{i+1} t^{p-2} dt = (i + 1)^{p-1} - i^{p-1} \leq \max(1, p - 1)(i + 1)^{p-2}.$$

Therefore

$$M_{p,\alpha}(Q) \leq \max(1, p - 1) \sum_{i=0}^\infty (i + 1)^{p-2} \int_0^{\alpha_i} Q^p(u) du \tag{C.9}$$

and

$$M_{p,\alpha,n}(Q) \leq \max(1, p - 1) \sum_{i=0}^{n-1} (i + 1)^{p-2} \int_0^{\alpha_i} Q^p(u) du. \tag{C.10}$$

We now apply (C.9) to random variables satisfying a tail assumption. Suppose that

$$\mathbb{P}(X > x) \leq (c/x)^r.$$

Then $Q(u) \leq cu^{-1/r}$, and consequently, if $r > p$,

$$M_{p,\alpha}(Q) \leq \frac{cr}{r-p} \max(1, p-1) \sum_{i=0}^{\infty} (i+1)^{p-2} \alpha_i^{1-p/r}. \qquad (C.11)$$

Hence $M_{p,\alpha}(Q)$ is finite as soon as there exists some $r > p$ such that

$$\mathbb{P}(X > x) \leq (c/x)^r \text{ and } \sum_{i=0}^{\infty} (i+1)^{p-2} \alpha_i^{1-p/r} < \infty. \qquad (C.12)$$

Assume now that $p > 2$. We have in view bounds for $M_{p,\alpha,n}(Q)$ in the case where $M_{p,\alpha}(Q) = \infty$. Recall that, by Theorem 6.3,

$$\mathbb{E}\left(\sup_{k\in[1,n]} |S_k|^p \right) \leq a_p s_n^p + n b_p M_{p,\alpha,n}(Q).$$

Hence, if $M_{p,\alpha,n}(Q) = O(n^q)$ for some $q \leq (p-2)/2$, the Marcinkiewicz–Zygmund inequality of order p holds true. Now, let $s < p-1$. From (C.10), the mixing condition

$$\int_0^{\alpha_i} Q^p(u) du = O((i+1)^{-s}) \qquad (C.13)$$

ensures that

$$n M_{p,\alpha,n}(Q) = O(n^{p-s}) \text{ as } n \to \infty. \qquad (C.14)$$

In particular, if $p > 2$ and $s = p/2$,

$$n M_{p,\alpha,n}(Q) = O(n^{p/2}) \text{ as soon as } \int_0^{\alpha_i} Q^p(u) du = O((i+1)^{-p/2}). \qquad (C.15)$$

In the case of bounded random variables, (C.15) holds true as soon as $\alpha_i = O(i^{-p/2})$. In the unbounded case, (C.15) holds true, for example, if there exists some $r > p$ such that

$$\mathbb{P}(X > x) \leq (c/x)^r \text{ and } \alpha_i = O(i^{-pr/(2r-2p)}). \qquad (C.16)$$

Geometric rates of mixing. Assume that $\alpha_i = O(a^i)$ for some $a < 1$. Then, using the same arguments as in the proof of (1.33), we get that $M_{p,\alpha}(Q)$ is finite as soon as

$$\mathbb{E}(X^p (\log(1+X))^{p-1}) < \infty. \qquad (C.17)$$

Annex D
Two Versions of a Lemma of Pisier

In this annex, we give an upper bound for the expectation of the maximum of a finite number of integrable random variables. This bound is then used to obtain upper bounds on this expectation under some assumptions on the Laplace transform or on the moments of the random variables in the style of Lemma 1.6 in Pisier (1983).

Proposition D.1 *Let $Z_1, \ldots, Z_N$ be a finite family of real-valued integrable random variables. Let F_i be the distribution function of Z_i. Let $F = F_1 + F_2 + \cdots + F_N$ and let F^{-1} denote the generalized inverse of F. Then*

$$\mathbb{E}\big(\max(Z_1, Z_2, \ldots, Z_N)\big) \leq \int_{N-1}^{N} F^{-1}(u)du.$$

Remark D.1 Let $H_i = 1 - F_i$ denote the tail function of X_i. Let $H = H_1 + H_2 + \cdots + H_N$ and let H^{-1} denote the generalized inverse of F. Then Proposition D.1 is equivalent to

$$\mathbb{E}\big(\max(Z_1, Z_2, \ldots, Z_N)\big) \leq \int_0^1 H^{-1}(u)du.$$

Proof For $N = 1$, Proposition D.1 is obvious. Let $N \geq 2$ and $T = \max(Z_1, Z_2, \ldots, Z_N)$. For any real t,

$$T \leq t + \sup_{i \in [1,N]} (Z_i - t)_+ \leq t + \sum_{i=1}^{N} (Z_i - t)_+. \tag{D.1}$$

Hence

$$\mathbb{E}(T) \leq t + \sum_{i=1}^{N} \mathbb{E}(Z_i - t)_+. \tag{D.2}$$

© Springer-Verlag GmbH Germany 2017
E. Rio, *Asymptotic Theory of Weakly Dependent Random Processes*,
Probability Theory and Stochastic Modelling 80,
DOI 10.1007/978-3-662-54323-8

Next

$$\sum_{i=1}^{N} \mathbb{E}(Z_i - t)_+ = \int_{\mathbb{R}} (z - t)_+ dF(z) = \int_0^N (F^{-1}(u) - t)_+ du.$$

Choosing $t = F^{-1}(N - 1)$ in the above formula, we obtain:

$$t + \sum_{i=1}^{N} \mathbb{E}(Z_i - t)_+ = F^{-1}(N - 1) + \int_{N-1}^N (F^{-1}(u) - F^{-1}(N - 1))du,$$

which implies Proposition D.1. ∎

Application to exponential tails. Assume that $H_i(t) \le \exp(-t)$ for any positive t and any i in $[1, N]$. Then $H(t) \le N \exp(-t)$, which ensures that $H^{-1}(x) \le \log(N/x)$. It follows that $\mathbb{E}(T) \le 1 + \log N$.

Application to power-type tails. Assume that $H_i(t) \le (a_i/t)^p$ for some $p > 1$ and some finite sequence $(a_i)_i$ of positive reals. Let $\|a\|_p = (a_1^p + \cdots + a_N^p)^{1/p}$. Then $H(t) \le (\|a\|_p/t)^p$, which ensures that $H^{-1}(u) \le \|a\|_p u^{-1/p}$. It follows that $(p - 1)\mathbb{E}(T) \le p\|a\|_p$.

We now give an application of Proposition D.1 to random variables with finite exponential moments. The lemma below is stated in Massart and Rio (1998). A short proof is given in Rio (1998).

Lemma D.1 *Let $(Z_i)_{i \in I}$ be a finite family of real-valued random variables. Suppose there exists some convex and nondecreasing function L, taking finite values on a right neighborhood of 0, such that $\log \mathbb{E}(\exp(tZ_i)) \le L(t)$ for any nonnegative t and any i in I. Let h_L be the Young transform of L and let H denote the logarithm of the cardinality of I. Then*

$$\mathbb{E}(\sup_{i \in I} Z_i) \le h_L^{-1}(H).$$

Proof We may assume that $I = \{1, 2, \ldots, N\}$. Let $T = \sup_{i \in I} Z_i$. By Proposition D.1 and Jensen's inequality, for any positive t,

$$\exp(t\mathbb{E}(T)) \le \int_{N-1}^N \exp(tF^{-1}(u))du \le \int_0^N \exp(tF^{-1}(u))du.$$

Now

$$\int_0^N \exp(tF^{-1}(u))du = \int_{\mathbb{R}} \exp(tx)dF(x) = \sum_{i=1}^{N} \mathbb{E}(\exp(tZ_i)) \le N \exp(L(t)).$$

Taking the logarithm, dividing by t and minimizing with respect to t, we infer that

$$\mathbb{E}(T) \le \inf_{t>0} t^{-1}(L(t) + H).$$

Lemma D.1 then follows from Lemma A.2. ∎

Application to exponential tails (continued). From the assumption, we may apply Lemma D.1 with the logarithm of the Laplace transform of the standard exponential law: Lemma D.1 holds with $L(x) = -\log(1-x)$ for $x \geq 0$. Then $L^*(t) = 0$ for $t \leq 1$ and $L^*(t) = t - 1 - \log t$ for $t \geq 1$. Consequently, Lemma D.1 yields $\mathbb{E}(T) \leq M$ with $M \geq 1$ the solution of the equation $M = 1 + \log(MN)$. Note that $M - 1 - \log N \geq \log(1 + \log N)$, which gives the order of the loss when applying Lemma D.1 instead of Proposition D.1.

We now consider random variables with finite moments.

Lemma D.2 *Let $(Z_i)_{i \in I}$ be a finite family of nonnegative real-valued random variables. Suppose there exists some convex and nondecreasing function M, taking finite values on a right neighborhood of 1, such that $\log \mathbb{E}(Z_i^r) \leq M(r)$ for any $r \geq 1$ and any i in I. Let h_M be the Young transform of M and H be the logarithm of the cardinality of I. Then*

$$\mathbb{E}(\sup_{i \in I} Z_i) \leq \exp(h_M^{-1}(H)).$$

Proof We may assume that $I = \{1, 2, \ldots, N\}$. Let $T = \sup_{i \in I} Z_i$. Starting from Proposition D.1 and applying Hölder's inequality, we get that

$$(\mathbb{E}(T))^r \leq \int_{N-1}^{N} (F^{-1}(u))^r du \leq \int_0^N (F^{-1}(u))^r du.$$

Now

$$\int_0^N (F^{-1}(u))^r du = \int_0^\infty x^r dF(x) = \sum_{i=1}^N \mathbb{E}(Z_i^r) \leq N \exp(M(r)).$$

Taking the logarithm, dividing by r and minimizing with respect to r, we infer that

$$\log \mathbb{E}(T) \leq \inf_{r \geq 1} r^{-1}(M(r) + H).$$

Lemma D.2 then follows from Lemma A.2. ∎

Annex E
Classical Results on Measurability

In this annex, we first recall a lemma of Skorohod (1976) on the representation of random variables. Next we give some properties of projections, which are helpful when proving the measurability of some functions (see Dellacherie 1972, Chap. 1). We first recall a lemma which may be found in Skorohod (1976).

Lemma E.1 *Let $\mathcal{X}$ be a Polish space. Then there exists a one to one mapping f from $\mathcal{X}$ onto a Borel subset of $[0, 1]$, which is bi-measurable with respect to the Borel σ-fields.*

Starting from Lemma E.1, we now prove a lemma of Skorohod (1976) stated below.

Lemma E.2 *Let $\mathcal{X}$ be a Polish space and let X be a random variable from $(\Omega, \mathcal{T}, \mathbb{P})$ into $\mathcal{X}$ equipped with its Borel σ-field $\mathcal{B}(\mathcal{X})$. Let $\mathcal{A}$ be a σ-field in $(\Omega, \mathcal{T}, \mathbb{P})$ and δ be a random variable with uniform distribution over $[0, 1]$, independent of $\mathcal{A} \vee \sigma(X)$. Then there exists a measurable mapping g from $(\Omega \times [0, 1], \mathcal{A} \otimes \mathcal{B}([0, 1]))$ into $\mathcal{X}$ and a random variable V with uniform law over $[0, 1]$, measurable with respect $\mathcal{A} \vee \sigma(X) \vee \sigma(\delta)$ and independent of $\mathcal{A}$ such that $X = g(\omega, V)$ almost surely.*

Proof From Lemma E.1, it is enough to prove Lemma E.2 in the case $\mathcal{X} = [0, 1]$. Let $F_{\mathcal{A}}(t) = \mathbb{P}(X \leq t \mid \mathcal{A})$ denote the conditional distribution function of X. Then the random variable

$$V = F_{\mathcal{A}}(X - 0) + \delta(F_{\mathcal{A}}(X) - F_{\mathcal{A}}(X - 0))$$

is measurable with respect to $\mathcal{A} \vee \sigma(X) \vee \sigma(\delta)$, independent of $\mathcal{A}$, and V has the uniform law over $[0, 1]$ (see Annex F). Now the mapping g defined by $g(\omega, v) = F_{\mathcal{A}}^{-1}(v)$ is measurable with respect to $\mathcal{A} \otimes \mathcal{B}([0, 1])$ and g satisfies $X = g(\omega, V)$, which completes the proof of Lemma E.2. ∎

We now recall a theorem of Dellacherie (1972, Theorem T32, p. 17) on projections.

© Springer-Verlag GmbH Germany 2017
E. Rio, *Asymptotic Theory of Weakly Dependent Random Processes*,
Probability Theory and Stochastic Modelling 80,
DOI 10.1007/978-3-662-54323-8

Theorem E.1 *Let $(\Omega, \mathcal{F}, P)$ be a complete probability space and $(K, \mathcal{B}(K))$ be a countably generated and locally compact space equipped with its Borel σ-field. Let us denote by π the canonical projection from $K \times \Omega$ on Ω. Then, for any B in $\mathcal{B}(K) \otimes \mathcal{F}$, the set $\pi(B)$ belongs to $\mathcal{F}$.*

We refer to Dudley (1989, Chap. 13) for more about the measurability properties of projections and for universally measurable sets and universally measurable functions, which are defined below.

Definition E.1 *Let $(X, \mathcal{X})$ be a measurable space and A be a subset of X. Then A is said to be universally measurable if, for any law P on $(X, \mathcal{X})$, A belongs to the completed σ-field of $\mathcal{X}$ for P. Let $(Y, \mathcal{Y})$ be a measurable space. A mapping f from X into Y is said to be universally measurable if, for any B in $\mathcal{Y}$, the set $f^{-1}(B)$ is universally measurable in X.*

To complete this section, we now give a slightly different formulation of Theorem E.1, using universally measurable sets.

Corollary E.1 *Let $(X, \mathcal{X})$ be a measurable space and let $(K, \mathcal{B}(K))$ be a countably generated and locally compact space equipped with its Borel σ-field. Let us denote by π the canonical projection from $K \times \Omega$ on Ω. Then, for any B in $\mathcal{B}(K) \otimes \mathcal{F}$, the set $\pi(B)$ is universally measurable.*

Annex F
The Conditional Quantile Transformation

In this annex, we will study the properties of the so-called conditional quantile transformation introduced in the proof of Lemma 5.2 and in the proof of Skorohod's lemma. The first step to define this transformation is to define a measurable selection of the conditional distribution function.

Let $\mathcal{A}$ be a σ-field in $(\Omega, \mathcal{T}, \mathbb{P})$ and X be a real-valued random variable. For any rational number q, we set

$$F_{\mathcal{A}}(q) = \mathbb{P}(X \leq q \mid \mathcal{A}).$$

The so-defined random function is almost surely defined on $\mathbf{Q}$, and this function is nondecreasing. The conditional distribution function is defined as the unique right-continuous function extending this function to the set of reals. Consequently, for any real x,

$$F_{\mathcal{A}}(x) = \lim_{\substack{q \searrow x \\ q \in Q}} F_{\mathcal{A}}(q). \tag{F.1}$$

The function $F_{\mathcal{A}}$ defined by (F.1) has the property below: the map which sends (x, ω) on $F_{\mathcal{A}}(x)$ is measurable with respect to the completed σ-field associated to $\mathcal{B}(\mathbb{R}) \otimes \mathcal{A}$.

We now define the conditional quantile transformation.

Lemma F.1 *Let X be a real-valued random variable, $\mathcal{A}$ be a σ-field of $(\Omega, \mathcal{T}, \mathbb{P})$ and δ be a random variable with uniform distribution over $[0, 1]$, independent of $\sigma(X) \vee \mathcal{A}$. Let $F_{\mathcal{A}}$ be the conditional distribution function defined by (F.1). Set*

$$V = F_{\mathcal{A}}(X - 0) + \delta(F_{\mathcal{A}}(X) - F_{\mathcal{A}}(X - 0)).$$

Then V has the uniform distribution over $[0, 1]$, and is independent of $\mathcal{A}$. Furthermore, $F_{\mathcal{A}}^{-1}(V) = X$ almost surely.

© Springer-Verlag GmbH Germany 2017

E. Rio, *Asymptotic Theory of Weakly Dependent Random Processes*,
Probability Theory and Stochastic Modelling 80,
DOI 10.1007/978-3-662-54323-8

Proof Let $v(\omega, x, t) = F_{\mathcal{A}}(x-0) + t(F_{\mathcal{A}}(x) - F_{\mathcal{A}}(x-0))$. The so-defined mapping v is measurable with respect to $\mathcal{A} \otimes \mathcal{B}(\mathbb{R}) \otimes \mathcal{B}(\mathbb{R})$. Hence $V = v(\omega, X, \delta)$ is a real-valued random variable. Let a be any real in $[0, 1]$. Let us consider

$$b = F_{\mathcal{A}}^{-1}(a + 0) = \sup\{x \in \mathbb{R} : F_{\mathcal{A}}(x) \le a\}.$$

If $F_{\mathcal{A}}$ is continuous at point b, then $F_{\mathcal{A}}(b) = a$. In that case $(v(\omega, x, t) \le a)$ if and only if $(x \le b)$, which ensures that $\mathbb{P}(V \le a \mid \mathcal{A}) = \mathbb{P}(X \le b \mid \mathcal{A}) = F_{\mathcal{A}}(b) = a$.

If $F_{\mathcal{A}}$ is not continuous at point b, then a belongs to $[F_{\mathcal{A}}(b-0), F_{\mathcal{A}}(b)]$, which implies that

$$a = v(\omega, b, u) \text{ for some } u \in [0, 1].$$

In that case, $(v(\omega, x, t) \le a)$ if and only if either $(x < b)$ or $(x = b$ and $t \le u)$. Then

$$\mathbb{P}(V \le a \mid \mathcal{A}) = F_{\mathcal{A}}(b-0) + u(F_{\mathcal{A}}(b) - F_{\mathcal{A}}(b-0)) = a.$$

Consequently, V has the uniform distribution over $[0, 1]$, conditionally to $\mathcal{A}$, and therefore V is uniformly distributed over $[0, 1]$.

Now, since x belongs to the set of reals y such that $F_{\mathcal{A}}(y) \ge v(\omega, x, t)$, we have:

$$x \ge F_{\mathcal{A}}^{-1}(v(\omega, x, t)) \text{ for any } t \in [0, 1].$$

It follows that $X \ge F_{\mathcal{A}}^{-1}(V)$ almost surely. Let ϕ be the distribution function of the standard normal law. Since $(F_{\mathcal{A}}^{-1}(V) > t)$ if and only if $(V > F_{\mathcal{A}}(t))$, we have:

$$\mathbb{E}(\phi(F_{\mathcal{A}}^{-1}(V)) \mid \mathcal{A}) = \int_{\mathbb{R}} \mathbb{P}(F_{\mathcal{A}}^{-1}(V) > t \mid \mathcal{A})\phi'(t)dt$$

$$= \int_{\mathbb{R}} \mathbb{P}(V > F_{\mathcal{A}}(t) \mid \mathcal{A})\phi'(t)dt$$

$$= \int_{\mathbb{R}} (1 - F_{\mathcal{A}}(t))\phi'(t)dt = \mathbb{E}(\phi(X) \mid \mathcal{A}).$$

It follows that $\mathbb{E}(\phi(X)) = \mathbb{E}(\phi(F_{\mathcal{A}}^{-1}(V)))$. Since $\phi(X) \ge \phi(F_{\mathcal{A}}^{-1}(V))$ almost surely, it implies that $\phi(X) = \phi(F_{\mathcal{A}}^{-1}(V))$ almost surely. Hence $X = F_{\mathcal{A}}^{-1}(V)$ almost surely, which completes the proof of Lemma F.1.

Annex G
Technical Tools

Lemma G.1 *For any nonnegative reals a and c,*

$$ac(c \wedge 1) \leq \tfrac{2}{3}c^2(c \wedge 1) + \tfrac{1}{2}a^2(a \wedge 1), \qquad (a)$$

$$a^2(c \wedge 1) \leq \tfrac{1}{3}c^2(c \wedge 1) + a^2(a \wedge 1). \qquad (b)$$

Proof of Lemma G.1 To prove (a), note that, if $a \leq 1$, then, by Young's inequality,

$$ac(c \wedge 1) = ac(c \wedge 1) \leq \tfrac{2}{3}c^{3/2}(c \wedge 1)^{3/2} + \tfrac{1}{3}a^3 \leq \tfrac{2}{3}c^2(c \wedge 1) + \tfrac{1}{3}a^2(a \wedge 1).$$

Then (a) holds. If $a \geq 1$, then

$$ac(c \wedge 1) \leq \leq (a^2 + c^2(c \wedge 1)^2)/2 \leq (a^2(a \wedge 1) + c^2(c \wedge 1))/2.$$

Consequently (a) still holds true.

We now prove (b). If $a \geq 1$, then $a^2(c \wedge 1) \leq a^2 \leq a^2(a \wedge 1)$ and (b) holds true. If $a \leq 1$, then, by Young's inequality,

$$a^2(c \wedge 1) \leq \tfrac{2}{3}a^3 + \tfrac{1}{3}(c \wedge 1)^3 \leq a^2(a \wedge 1) + \tfrac{1}{3}c^2(c \wedge 1),$$

which completes the proof of (b). ∎

© Springer-Verlag GmbH Germany 2017
E. Rio, *Asymptotic Theory of Weakly Dependent Random Processes*,
Probability Theory and Stochastic Modelling 80,
DOI 10.1007/978-3-662-54323-8

References

Andersen, N. T., Giné, E., Ossiander, M. and Zinn, J. (1988). The central limit theorem and the law of iterated logarithm for empirical processes under local conditions. *Probab. Th. Rel. Fields* **77**, 271–305.

Ango-Nzé, P. (1994). Critères d'ergodicité de modèles markoviens. Estimation non paramétrique sous des hypothèses de dépendance. *Thèse de doctorat d'université. Université Paris 9, Dauphine.*

Arcones, M. A. and Yu, B. (1994). Central limit theorems for empirical and U-processes of stationary mixing sequences. *J. Theoret. Prob.* **7**, 47–71.

Azuma, K. (1967). Weighted sums of certain dependent random variables. *Tôkohu Math. J.* **19**, 357–367.

Bártfai, P. (1970). Über die Entfernung der Irrfahrtswege. *Studia Sci. Math. Hungar.* **1**, 161–168.

Bass, J. (1955). Sur la compatibilité des fonctions de répartition. *C.R. Acad. Sci. Paris* **240**, 839–841.

Berbee, H. (1979). Random walks with stationary increments and renewal theory. *Math. Cent. Tracts, Amsterdam.*

Berbee, H. (1987). Convergence rates in the strong law for bounded mixing sequences. *Probab. Th. Rel. Fields* **74**, 255–270.

Berkes, I. and Philipp, W. (1979). Approximation theorems for independent and weakly dependent random vectors. *Ann. Probab.* **7**, 29–54.

Bercu, B., Delyon, B. and Rio, E. (2015). Concentration inequalities for sums and martingales. *Springer Briefs in Mathematics.* Springer, Heidelberg.

Bergström, H. (1972). On the convergence of sums of random variables in distribution under mixing condition. Collection of articles dedicated to the memory of Alfred Rényi, I. *Period. Math. Hungar.* **2**, 173–190.

Billingsley, P. (1968). Convergence of probability measures. Wiley, New-York.

Billingsley, P. (1985). Probability and Measure. Second edition. Wiley, New-York.

Bolthausen, E. (1980). The Berry-Esseen theorem for functionals of discrete Markov chains. *Z. Wahrsch. verw. Gebiete* **54**, 59–73.

Bolthausen, E. (1982a). On the central limit theorem for stationary mixing random fields. *Ann. Probab.* **10**, 1047–1050.

Bolthausen, E. (1982b). The Berry-Esseen theorem for strongly mixing Harris recurrent Markov chains. *Z. Wahrsch. verw. Gebiete* **60**, 283–289.

Bosq, D. (1993). Bernstein's type large deviation inequalities for partial sums of strong mixing process. *Statistics* **24**, 59–70.

© Springer-Verlag GmbH Germany 2017

E. Rio, *Asymptotic Theory of Weakly Dependent Random Processes*, Probability Theory and Stochastic Modelling 80, DOI 10.1007/978-3-662-54323-8

Bradley, R. C. (1983). Approximation theorems for strongly mixing random variables. *Michigan Math. J.* **30**, 69–81.

Bradley, R. C. (1986). Basic properties of strong mixing conditions. *Dependence in probability and statistics. A survey of recent results. Oberwolfach, 1985.* E. Eberlein and M.S. Taqqu editors. Birkhäuser.

Bradley, R. C. (1997). On quantiles and the central limit question for strongly mixing sequences. *J. of Theor. Probab.* **10**, 507–555.

Bradley, R. C. (2005). Basic properties of strong mixing conditions. A survey and some open questions. Update of, and a supplement to, the 1986 original. *Probab. Surv.* **2**, 107–144.

Bradley, R. C. (2007). Introduction to strong mixing conditions. Volumes 1, 2 and 3. Kendrick Press, Hebert City (Utah).

Bradley, R. C. and Bryc, W. (1985). Multilinear forms and measures of dependence between random variables. *J. Multivar. Anal.* **16**, 335–367.

Bretagnolle, J. and Huber, C. (1979). Estimation des densités: risque minimax. *Z. Wahrsch. verw. Gebiete* **47**, 119–137.

Bücher, A. (2015). A note on weak convergence of the sequential multivariate empirical process under strong mixing. *J. Theoret. Probab.* **28**, no. 3, 1028–1037.

Bulinskii, A. and Doukhan P. (1987). Inégalités de mélange fort utilisant des normes d'Orlicz. *C.R. Acad. Sci. Paris, Série I.* **305**, 827–830.

Collomb, G. (1984). Propriétés de convergence presque complète du prédicteur à noyau. *Z. Wahrsch. verw. Gebiete* **66**, 441–460.

Cuny, C. and Fan, A.H. (2016). Study of almost everywhere convergence of series by means of martingale methods. *Preprint.*

Davydov, Y. A., (1968). Convergence of distributions generated by stationary stochastic processes. *Theor. Probab. Appl.* **13**, 691–696.

Davydov, Y. A. (1973). Mixing conditions for Markov chains. *Theor. Probab. Appl.* **18**, 312–328.

Dedecker, J. (2004). Inégalités de covariance. *C. R. Math. Acad. Sci. Paris* **339**, no. 7, 503–506.

Dedecker, J. (2010). An empirical central limit theorem for intermittent maps. *Probab. Theory Related Fields* **148**, no. 1–2, 177–195. Erratum, *Probab. Theory Related Fields* **155** (2013), no. 1–2, 487–491.

Dedecker, J. and Doukhan, P. (2003). A new covariance inequality and applications. *Stochastic Process. Appl.* **106**, no. 1, 63–80.

Dedecker, J., Doukhan, P., Lang, G., León, J. R., Louhichi, S., Prieur, C. (2007). Weak dependence: with examples and applications. *Lecture Notes in Statistics* **190**. Springer, New York.

Dedecker, J., Gouëzel, S. and Merlevède, F. (2010). Some almost sure results for unbounded functions of intermittent maps and their associated Markov chains. *Ann. Inst. Henri Poincar Probab. Stat.* **46**, no. 3, 796–821.

Dedecker, J. and Merlevède, F. (2007). Convergence rates in the law of large numbers for Banach-valued dependent variables. *Teor. Veroyatn. Primen.* **52**, no. 3, 562–587.

Dedecker, J. and Prieur, C. (2004). Coupling for τ-dependent sequences and applications. *J. Theoret. Probab.* **17**, no. 4, 861–885.

Dedecker, J. and Prieur, C. (2005). New dependence coefficients. Examples and applications to statistics. *Probab. Theory Related Fields* **132**, no. 2, 203–236.

Dehling, H. (1983). Limit theorems for sums of weakly dependent Banach space valued random variables. *Z. Wahrsch. Verw. Gebiete* **63**, no. 3, 393–432.

Dehling, H., Durieu, O. and Volný, D. (2009). New techniques for empirical processes of dependent data. *Stochastic Process. Appl.* **119**, no. 10, 3699–3718.

Dehling, H., Mikosch, T. and Sørensen, M. (eds) (2002). Empirical Processes Techniques for Dependent Data. Birkhäuser, Boston.

Dellacherie, C. (1972). Capacités et processus stochastiques. *Ergebnisse der mathematik und ihrer grenzgebiete.* Springer, Berlin.

Dellacherie, C. and Meyer, P. A. (1975). Probabilité et potentiel. Masson. Paris.

Delyon, B. (1990). Limit theorem for mixing processes. *Tech. Report IRISA, Rennes 1,* **546**.

Delyon, B. (2015). Exponential inequalities for dependent processes. hal-01072019.

Devore, R. A. and Lorentz, G. G. (1993). Constructive approximation. *Die grundlehren der mathematischen wissenschaften.* Springer, Berlin.

Dhompongsa, S. (1984). A note on the almost sure approximation of empirical process of weakly dependent random vectors. *Yokohama math. J.* **32**, 113–121.

Doeblin, W. (1938). Sur les propriétés asymptotiques de mouvements régis par certains types de chaînes simples. *Thèse de doctorat.*

Donsker, M. (1952). Justification and extension of Doob's heuristic approach to the Kolmogorov–Smirnov's theorems. *Ann. Math. Stat.* **23**, 277–281.

Doukhan, P. (1994). Mixing: properties and examples. *Lecture Notes in Statistics* **85**. Springer, New-York.

Doukhan, P. and Ghindès, M. (1983). Estimation de la transition de probabilité d'une chaîne de Markov Doeblin-récurrente. Étude du cas du processus autoégressif général d'ordre 1. *Stochastic processes appl.* **15**, 271–293.

Doukhan, P., León, J. and Portal, F. (1984). Vitesse de convergence dans le théorème central limite pour des variables aléatoires mélangeantes à valeurs dans un espace de Hilbert. *C. R. Acad. Sci. Paris Série 1.* **298**, 305–308.

Doukhan, P., León, J. and Portal, F. (1987). Principe d'invariance faible pour la mesure empirique d'une suite de variables aléatoires dépendantes. *Probab. Th. Rel. Fields* **76**, 51–70.

Doukhan, P., Massart, P. and Rio, E. (1994). The functional central limit theorem for strongly mixing processes. *Annales inst. H. Poincaré Probab. Statist.* **30**, 63–82.

Doukhan, P., Massart, P. and Rio, E. (1995). Invariance principles for absolutely regular empirical processes. *Ann. Inst. H. Poincaré Probab. Statist.* **31**, 393–427.

Doukhan, P. and Portal, F. (1983). Moments de variables aléatoires mélangeantes. *C. R. Acad. Sci. Paris Série 1.* **297**, 129–132.

Doukhan, P. and Portal, F. (1987). Principe d'invariance faible pour la fonction de répartition empirique dans un cadre multidimensionnel et mélangeant. *Probab. math. statist.* **8–2**, 117–132.

Doukhan, P. and Surgailis, D. (1998). Functional central limit theorem for the empirical process of short memory linear processes. *C. R. Acad. Sci. Paris Sr. I Math.* **326**, no. 1, 87–92.

Dudley, R. M. (1966). Weak convergence of probabilities on nonseparable metric spaces and empirical measures on Euclidean spaces. *Illinois J. Math.* **10**, 109–126.

Dudley, R. M. (1967). The sizes of compact subsets of Hilbert space and continuity of Gaussian processes. *J. Functional Analysis* **1**, 290–330.

Dudley, R. M. (1978). Central limit theorems for empirical measures. *Ann. Probab.* **6**, 899–929.

Dudley, R. M. (1984). A course on empirical processes. Ecole d'été de probabilités de Saint-Flour XII-1982. *Lectures Notes in Math.* **1097**, 1–142. Springer. Berlin.

Dudley, R. M. (1989). Real analysis and probability. Wadsworth Inc., Belmont, California.

Feller, W. (1950). An introduction to probability theory and its applications. Wiley, New-York.

Fréchet, M., (1951). Sur les tableaux de corrélation dont les marges sont données. *Annales de l'université de Lyon, Sciences, section A.* **14**, 53–77.

Fréchet, M., (1957). Sur la distance de deux lois de probabilité. *C.R. Acad. Sci. Paris.* **244–6**, 689–692.

Fuk, D. Kh. and Nagaev, S. V. (1971). Probability inequalities for sums of independent random variables. *Theory Probab. Appl.* **16**, 643–660.

Garsia, A. M. (1965). A simple proof of E. Hopf's maximal ergodic theorem. *J. of Maths. and Mech.* **14**, 381–382.

Goldstein, S. (1979). Maximal coupling. *Z. Wahrsch. verw. Gebiete* **46**, 193–204.

Gordin, M. I. (1969). The central limit theorem for stationary processes. *Soviet Math. Dokl.* **10**, 1174–1176.

Gordin, M. I. (1973). Abstracts of Communications, T1: A-K. International Conference on Probability Theory at Vilnius, 1973.

Gordin, M. I. and Peligrad, M. (2011). On the functional central limit theorem via martingale approximation. *Bernoulli* **17**, no. 1, 424–440.

Griffeath, D. (1975). A maximal coupling for Markov chains. *Z. Wahrscheinlichkeitstheorie und Verw. Gebiete* **31**, 95–106.

Hall, P. and Heyde, C. C. (1980). Martingale limit theory and its application. North-Holland, New-York.

Herrndorf, N. (1985). A functional central limit theorem for strongly mixing sequences of random variables. *Z. Wahr. Verv. Gebiete* **69**, 541–550.

Ibragimov, I. A. (1962). Some limit theorems for stationary processes. *Theor. Probab. Appl.* **7**, 349–382.

Ibragimov, I. A. and Linnik, Y. V. (1971). Independent and stationary sequences of random variables. Wolters-Noordhoff, Amsterdam.

Jain, N. and Jamison, B. (1967). Contributions to Doeblin's theory of Markov processes. *Z. Wahr. Verv. Gebiete* **8**, 19–40.

Krieger, H. A. (1984). A new look at Bergström's theorem on convergence in distribution for sums of dependent random variables. *Israel J. Math.* **47**, no. 1, 32–64.

Leblanc, F. (1995). Estimation par ondelettes de la densité marginale d'un processus stochastique: temps discret, temps continu et discrétisation. *Thèse de doctorat d'université. Université Paris 6, Jussieu.*

Lindeberg, J. W. (1922). Eine neue Herleitung des Exponentialgezetzes in der Wahrscheinlichkeitsrechnung. *Mathematische Zeitschrift.* **15** 211–225.

Lindvall, T. (1979). On coupling of discrete renewal processes *Z. Wahr. Verv. Gebiete* **48**, 57–70.

Louhichi, S. (2000). Weak convergence for empirical processes of associated sequences. *Ann. Inst. H. Poincar Probab. Statist.* **36**, no. 5, 547–567.

Major, P. (1978). On the invariance principle for sums of identically distributed random variables. *J. Multivariate Anal.* **8**, 487–517.

Massart, P. (1987). Quelques problèmes de convergence pour des processus empiriques. *Doctorat d'état. Université Paris-Sud, centre d'Orsay.*

Massart, P. and Rio, E. (1998). A uniform Marcinkiewicz-Zygmund strong law of large numbers for empirical processes. *Asymptotic methods in probability and statistics (Ottawa, ON, 1997),* 199–211. North-Holland, Amsterdam.

Melbourne, I. and Nicol, M. (2008). Large deviations for nonuniformly hyperbolic systems. *Trans. Amer. Math. Soc.* 360, no. 12, 6661–6676.

Merlevède, F. and Peligrad, M. (2000). The functional central limit theorem under the strong mixing condition. *Ann. Probab.* **28**, no. 3, 1336–1352.

Merlevède, F., Peligrad, M. and Rio, E. (2011). A Bernstein type inequality and moderate deviations for weakly dependent sequences. *Probab. Theory Related Fields* **151**, no. 3–4, 435–474.

Merlevède, F., Peligrad, M. and Utev, S. (2006). Recent advances in invariance principles for stationary sequences. *Probab. Surv.* **3**, 1–36.

Meyer, Y. (1990). Ondelettes et opérateurs. Hermann, Paris.

Meyn, S. P. and Tweedie, R. L. (1993). Markov chains and stochastic stability. *Communications and control engineering series.* Springer, Berlin.

Mokkadem, A. (1985). Le modèle non linéaire AR(1) général. Ergocicité et ergodicité géométrique. *C.R. Acad. Sci. Paris Série 1.* **301**, 889–892.

Mokkadem, A. (1987). Critères de mélange pour des processus stationnaires. Estimation sous des hypothèses de mélange. Entropie des processus linéaires. *Doctorat d'état. Université Paris-Sud, centre d'Orsay.*

Neumann, M. H. (2013). A central limit theorem for triangular arrays of weakly dependent random variables, with applications in statistics. *ESAIM Probab. Stat.* **17**, 120–134.

Nummelin, E. (1978). A splitting technique for Harris recurrent Markov chains. *Z. Wahr. Verv. Gebiete* **43**, 309–318.

Nummelin, E. (1984). General irreducible Markov chains and non-negative operators. Cambridge University Press, London.

Oodaira, H. and Yoshihara, K. I. (1971a). The law of the iterated logarithm for stationary processes satisfying mixing conditions. *Kodai Math. Sem. Rep.* **23**, 311–334.

Oodaira, H. and Yoshihara, K. I. (1971b). Note on the law of the iterated logarithm for stationary processes satisfying mixing conditions. *Kodai Math. Sem. Rep.* **23**, 335–342.

Oodaira, H. and Yoshihara, K. I. (1972). Functional central limit theorems for strictly stationary processes satisfying the strong mixing condition. *Kodai Math. Sem. Rep.* **24**, 259–269.

Orey, S. (1971). Lecture notes on limit theorems for Markov chain transition. Van Nostrand, London.

Ossiander, M. (1987). A central limit theorem under metric entropy with L^2-bracketing. *Ann. Probab.* **15**, 897–919. (1987)

Peligrad, M. (1983). A note on two measures of dependence and mixing sequences. *Adv. Appl. Probab.* **15**, 461–464.

Peligrad, M. (2002). Some remarks on coupling of dependent random variables. *Statist. Probab. Lett.* **60**, no. 2, 201–209.

Petrov, V. V. (1989). Some inequalities for moments of sums of independent random variables. *Prob. Theory and Math. Stat. Fifth Vilnius conference.* VSP-Mokslas.

Pinelis, I. (1994). Optimum bounds for the distributions of martingales in Banach spaces. *Ann. Probab.* **22**, no. 4, 1679–1706.

Pisier, G. (1983). Some applications of the metric entropy condition to harmonic analysis. Banach spaces, harmonic analysis, and probability theory (Storrs, Conn., 1980/1981), 123–154, *Lecture Notes in Math.,* **995**. Springer, Berlin.

Pitman, J. W. (1974). Uniform rates of convergence for Markov chain transition. *Z. Wahr. Verv. Gebiete* **29**, 193–227.

Pollard, D. (1982). A central limit theorem for empirical processes. *J. Aust. Math. Soc. Ser. A* **33**, 235–248.

Pollard, D. (1984). Convergence of stochastic processes. Springer, Berlin.

Pollard, D. (1990). Empirical processes: theory and applications. *NSF-CBMS Regional Conference Series in Probability and Statistics.* IMS-ASA, Hayward-Alexandria.

Rio, E. (1993). Covariance inequalities for strongly mixing processes. *Ann. Inst. H. Poincaré Probab. Statist.* **29**, 587–597.

Rio, E. (1994). Local invariance principles and their application to density estimation. *Probab. Theory Related Fields* **98**, no. 1, 21–45.

Rio, E. (1995a). A maximal inequality and dependent Marcinkiewicz-Zygmund strong laws. *Ann. Probab.* **23**, 918–937.

Rio, E. (1995b). The functional law of the iterated logarithm for stationary strongly mixing sequences. *Ann. Probab.* **23**, 1188–1203.

Rio, E. (1995c). About the Lindeberg method for strongly mixing sequences. *ESAIM, Probabilités et Statistiques,* **1**, 35–61.

Rio, E. (1998). Processus empiriques absolument réguliers et entropie universelle. *Prob. Th. Rel. Fields,* **111**, 585–608.

Rio, E. (2009). Moment inequalities for sums of dependent random variables under projective conditions. *J. Theoret. Probab.* **22**, no. 1, 146–163.

Rosenblatt, M. (1956). A central limit theorem and a strong mixing condition. *Proc. Nat. Acad. Sci. U.S.A.* **42**, 43–47.

Rosenthal, H. P., (1970). On the subspaces of L^p (p > 2) spanned by sequences of independent random variables. *Israel J. Math.* **8**, 273–303.

Rozanov, Y. A. and Volkonskii, V. A. (1959). Some limit theorems for random functions I. *Theory Probab. Appl.* **4**, 178–197.

Serfling, R. J. (1970). Moment inequalities for the maximum cumulative sum. *Ann. Math. Statist.* **41**, 1227–1234.

Shao, Q-M. (1993). Complete convergence for α-mixing sequences. *Statist. Probab. Letters* **16**, 279–287.

Shao, Q-M. and Yu, H. (1996). Weak convergence for weighted empirical processes of dependent sequences. *Ann. Probab.* **24**, 2098–2127.

Skorohod, A. V. (1976). On a representation of random variables. *Theory Probab. Appl.* **21**, 628–632.

Stein, C. (1972). A bound on the error in the normal approximation to the distribution of a sum of dependent random variables. *Proc. Sixth Berkeley Symp. Math. Statist. and Prob.*, II, 583–602. Cambridge University Press, London.

Stone, C. and Wainger, S. (1967). One-sided error estimates in renewal theory. *Journal d'analyse mathématique* **20**, 325–352.

Stout, W. F. (1974). Almost sure convergence. Academic Press, New York.

Strassen, V. (1964). An invariance principle for the law of the iterated logarithm. *Z. Wahr. Verv. Gebiete* **3**, 211–226.

Szarek, S. (1976). On the best constants in the Khinchin inequality. *Studia Math.* **58**, 197–208.

Tuominen, P. and Tweedie, R. L. (1994). Subgeometric rates of convergence of f-ergodic Markov chains. *Adv. Appl. Prob.* **26**, 775–798.

Tweedie, R. L. (1974). R-theory for Markov chains on a general state space I: solidarity properties and R-recurrent chains. *Ann. Probab.* **2**, 840–864.

Ueno, T. (1960). On recurrent Markov processes. *Kodai Math. J. Sem. Rep.* **12**, 109–142.

Utev, S. (1985). Inequalities and estimates of the convergence rate for the weakly dependent case. In *Adv. in Probab. Th., Novosibirsk.*

Viennet, G. (1997). Inequalities for absolutely regular sequences: application to density estimation. *Prob. Th. Rel. Fields* **107**, 467–492.

Volný, D. (1993). Approximating martingales and the central limit theorem for strictly stationary processes. *Stochastic processes appl.* **44**, 41–74.

Wintenberger, O. (2010). Deviation inequalities for sums of weakly dependent time series. *Electron. Commun. Probab.* **15**, 489–503.

Yokoyama, R. (1980). Moment bounds for stationary mixing sequences. *Z. Wahr. Verv. Gebiete* **52**, 45–57.

Yoshihara, K. (1979). Note on an almost sure invariance principle for some empirical processes. *Yokohama math. J.* **27**, 105–110.

Index

A

Almost sure convergence of series, 52
Azuma inequality, 41

B

Bennett inequality, 101, 175
Bernstein inequality, 176
β-mixing coefficient, 22, 23

C

Càdlàg inverse function, 5
Central limit theorem for β-mixing sequences, 87
Central limit theorem for partial sums, 65
Communication structure, 150
Condition (DMR), 11–13
Conditional distribution function, 90
Conditional quantile transformation, 90, 193
Coupling lemma, 89, 91

D

Delyon's covariance inequality, 22, 24
Dispersion function D_F, 27
Distribution function, 5
Doeblin's condition, 159

E

Empirical distribution function, 114, 118
Empirical measure, 15
Entropy with bracketing, 131
Ergodicity criterion of Tuominen and Tweedie, 163
Even algebraic moments, 37

Exponential inequality, 40, 48, 101, 175

F

Fidi convergence, 117
Finite dimensional convergence, 117
Fourth moment of sums, 33
(f, ψ)-ergodicity, 163
Fuk-Nagaev inequality, 103
Functional central limit theorem, 68

G

Gaussian process, 117
Geometric and subgeometric rates of mixing, 13

H

Histogram bases, 20
Hoeffding's inequality, 41

I

Incidence process, 153
Inequality of Hoeffding, 179
Irreducibility measures, 151
Irreducible stochastic kernel, 151

K

Kernel density estimators, 15

L

Law of the iterated logarithm, 109, 111
Lemma of Skorohod, 191

© Springer-Verlag GmbH Germany 2017
E. Rio, *Asymptotic Theory of Weakly Dependent Random Processes*,
Probability Theory and Stochastic Modelling 80,
DOI 10.1007/978-3-662-54323-8

Lindeberg condition, 71
Luxemburg norm, 173

M
Marcinkiewicz-Zygmund type inequality, 45, 107
Maximal coupling, 97
Maximal inequality of Kolmogorov, 51
Maximal irreducibility measure, 151
Mean integrated square error, 20
Metric entropy, 131
Moment inequalities, 106
Multivariate distribution functions, 121

O
Orlicz spaces, 173

P
Pitman occupation measure, 155
Positive kernel, 150
Potential, 150
Projection estimator, 16, 26

Q
Quantile function, 5

R
Rademacher sequence, 18
Rates of strong mixing and rates of ergodicity, 164
Recurrence, 153
Renewal process, 152
Riesz bases, 21

S
Series of covariances, 2, 3, 10, 66
Small sets, 159
Stationarity at second order, 1
Stein method, 84
Stochastic equicontinuity, 117, 121
Strict stationarity, 1
Strong law, 54, 55
Strong mixing coefficient, 3, 4

T
Transition probability kernel, 150
Triangular arrays, 70

U
Unconditional basis, 20
Uniform central limit theorem, 116, 117
Uniform central limit theorem of Pollard, 145
Universal entropy, 145

V
Variance of partial sums, 9
φ-mixing coefficient, 22

W
Weighted moments, 10, 183

Y
Young dual function, 171

Printed in the United States
By Bookmasters